Liquid–Liquid Extraction

Liquid–Liquid Extraction

RUTH BLUMBERG

1988

ACADEMIC PRESS

Harcourt Brace Jovanovich, Publishers
London   San Diego   New York
Boston   Sydney   Tokyo   Toronto

ACADEMIC PRESS LIMITED
24–28 Oval Road
London NW1

United States Edition published by
ACADEMIC PRESS INC.
San Diego, CA 92101

Copyright © 1988 by
ACADEMIC PRESS LIMITED

All Rights Reserved
No part of this book may be reproduced in any form, by photostat, microfilm or any other means, without written permission from the publishers

ISBN 0–12–107820–5

British Library Cataloguing in Publication Data
is available

Phototypeset by Paston Press, Loddon, Norfolk
and printed in Great Britain by
St Edmundsbury Press Limited, Bury St Edmunds, Suffolk

Preface

This monograph is intended for those already working in the liquid–liquid extraction (LLX) field. It is not a formal teaching manual, nor is it a review of solvent extraction processes. It is an exposé of an integrated approach to the development and design of liquid–liquid extraction processes, based on broad personal experience, and on a generalized research and development philosophy which extends to the development of any novel chemical process.

In this approach to research and development (R & D) the separate aspects which may have a bearing on the process are to be identified from the outset, as far as this is possible, and are then to be pursued in parallel, so that an outline of the whole is pictured from the beginning and becomes more clearly defined and tangible as work proceeds. It has been the aim, in writing this monograph, to show up as many as possible of the separate aspects of research which must be taken together to form the basis for the development of a liquid–liquid extraction process. The "thinking" models presented are based firmly on practice, experience and experimentation.

Liquid–liquid extraction with its unique flexibility is a powerful separation tool; the successful choice of the second liquid phase for an entirely new application of liquid–liquid extraction is an exciting, satisfying experience; not less so is the development of a workable process scheme based on this choice.

The skill of process development (PD), in its broadest sense, can be extended only "on the job", by working with others who have expertise and experience. The writing of this monograph is a welcome opportunity to place broad personal expertise and experience at the service of the wide spectrum of researchers who aim to put liquid–liquid extraction to use for novel purposes by developing new, viable processes.

Two central themes have been stressed—one is the great variety of two-phase systems that potentially exist, the other is the integrated approach in selecting and using LLX systems for practical purposes. Overall the integrated approach aims at asking relevant questions so that correct answers may be sought and found.

The text is divided into three parts—Part I exposes problems and questions relating to the scope of, and the approach to, the LLX procedure,

Part II concentrates on the LLX separation, Part III deals with the separation process itself. The overall approach is process oriented since the process is the ultimate aim and is the measure of R & D success. Examples and procedures are given in so far as these make the approach clearer; the examples are all drawn from personal experience. However, it is not the main purpose of the book to teach how "to do" LLX, but how "to think" LLX, so as to develop and expand its application and use.

Ruth Blumberg
Haifa, Israel

Acknowledgements

This book is the culmination of a life of striving, the odds being often against, sometimes for. Acknowledgements must therefore recognize all those who helped to make the way passable. In chronological order:
— To my late Parents, representing my line of hereditary and especially to my early-widowed Mother for her humane teachings, and her realizing that "education is the only lasting asset that a parent can confer on a child"
— To my first Employer in South Africa, who told me of Columbia University in the City of New York, and who urged me to go there to extend my studies—"if you want to get on professionally", as he said
— To my late Uncle, who paid my fare to the USA, and to my dear Sister and Brother-in-law, who took our family burden on themselves thus leaving me free to go
— To Columbia University Graduate School and the Chemical Engineering Department which awarded me a Graduate Residence Scholarship on faith and with goodwill
— To Professor T. B. Drew, then Head of the Chemical Engineering Department, who urged me to make up deficiencies and to go directly for a Ph.D. in Chemical Engineering
— To the late Dr. J. Blumenfeld, to Mr. E. Bodankin and Dr. A. M. Baniel, all of Israel Mining Industries, who said "come to us, to our Institute for Research and Development in Haifa"
— To Dr. Baniel, who first suggested we look at LLX, and all the colleagues at IMI with whom we built up our collective and individual reputations in this field
— To Dr. O. G. Wegrich of Miles Laboratories, Inc., USA, who offered me a rare framework for "doing and training" so that both Miles and I could benefit from my expertise without requiring that I relocate and leave my country.

More specifically connected with this book, thanks are extended:—
— To Peter Brown, then Editorial Director of Academic Press, who asked me to write it

Acknowledgements

— To Rodney Saddler, then Vice President Citric Operations of Miles, who left me free to do the writing in Israel and to the Miles management who understood that this is a positive contribution to the company
— To the Chemical Engineering Department of the Technion-Israel Institute of Technology, in particular to Ze'ev Tadmor then Dean of the Department and to the successive Deans, who made the writing pleasant by providing office space and library facilities as a "guest of the faculty"
— To my longtime colleague and close friend Alex Alon of Haifa, who read the earliest version of the text and encouraged me by his comments
— To Tom Foley, at the time Head of Citric Process Development at Miles, who offered to supervise the Word Processor and who suggested using Computer Graphics, actually doing part of the drawings himself; without him, the task would have been inordinately more complex and time-consuming. For detailed work, my thanks to Don Hess of Foley's Department who did most of the drawings, and to Foley's secretary Cathy Cooper who typed the text
— To the late Carl Hanson of Bradford University, U.K., who considered that I had "novel, interesting things to say about LLX", and to my many LLX friends particularly those who selected me for the ISEC '83 award for "outstanding accomplishment and meritorious achievement in solvent extraction"
— To my many colleagues in Biotechnology R & D at Miles, who have helped me to make the necessary adjustments in "thinking" when going over from Inorganic Processing to problems in Biotechnology
— To my good friends in Haifa, in particular Benjamin and Ora Cohen and more recently also David and Sarah Meyer, who have watched over my apartment and my affairs, thus enabling me to make the frequent visits to Miles Laboratories, Inc. in the USA with an easy mind
— To the many whom I have met under various circumstances and in widely different places, with whom ideas were discussed and approaches developed

Contents

Preface	v
Acknowledgements	vii
Introductory	xiii
Scope and Approach	xv

I   The Liquid–Liquid Extraction (LLX) Procedure— An Integrated Approach   1

Definition of the separation	1
Liquid–liquid extraction. Its nature and scope	12
Solvent components; extractant (reagent), diluent, modifier	15
Extractant characteristics; types, interactions, specificity, selectivity	17
Solvent selection	18
System properties and their significance	20
The shape of the equilibrium curve and its implications	23
Dilute solutions	26
Flowsheet delineation	28
The two-liquid phase system	32
The phase rule in LLX	33
Diluent/extractant interactions	36
Multi-component multi-phase systems in LLX	37
The second liquid phase—freedom of choice	41
In situ self-generation of two liquid phases	43
LLX as tool for achieving transformations	44
Extension of LLX—where and how?	46
Biotechnology	49
LLX in conjunction with fermentation	55
Envisaging new process applications for LLX	56
Systematization of solvent systems	61

	Technological aspects	63
	Economics/engineering of LLX operation	66
	Equipment characterization/basic aspects in relation to equipment selection	68
	Environmental interactions	69
	Peripheral operations/interactions with LLX step	70
	Process feasibility, proof vs degree of certainty	72
	LLX detriments	73
II	**The LLX Separation—Chemical Feasibility**	78
	LLX—various viewpoints	78
	Design of LLX systems	81
	LLX strategy and constraints	85
	Modes of studying LLX systems—specific approaches	89
	How to approach a LLX process study	93
	Laboratory simulation—process study of a defined system	99
	An exercise in developing an interdisciplinary program of study aimed at utilizing third-phase formation	103
	Interactive solvent systems	105
	Technological aspects of transfer from liquid to liquid	107
	LLX internal recycle/reflux—extension of the rectification/stripping analogy	120
	Computer simulations in LLX	126
	LLX—process synthesis—computer-aided process design	127
	Process synthesis (CAPD) generalized models—for what?	129
	Transfer modes	130
	Typical examples	133
	Reactions in LL systems	137
	Metal and acid extraction	141
III	**The LLX Separation Process—Technological Feasibility**	145
	Verification of the LLX separation process	145
	Process development	152
	Equipment and process	157
	Testing, scale, validity, piloting, etc.	160
	Equipment and process interactions	161
	Coalescence and entrainment	163
	Demixing/second- or third-liquid phase formation—coalescence and entrainment	166
	Clean-up in a LLX system	167
	Solids in LLX systems	169
	Process control	169

Inventory of surge in LLX systems 173
Evaluation of alternatives. Optimization vs timing, safety,
 economics 175
Re-evaluation with experience 177
Comparison of LLX with other procedures for achieving similar
 separation 180
Evaluation of LLX 182

Recommended Reading 184

Index 187

Dedication

This book is in memory of those young members of our family who were eliminated from this life in the 1940's and were thus ever prevented from encountering life's challenges and triumphs.

Introductory

One can assume that LLX for its own sake has no justification, since it is only a separation procedure, which, for success, requires that it be incorporated into a separation process. This means that the overriding aspect of an integrated LLX program will be the interactions and interconnections that enable the procedure to become the process. The presentation here aims to indicate how to make these interconnections.

The development of any process must be an inter-disciplinary activity, and so indeed the development of a LLX process depends on the interaction of many participants. The primary aims here have been to show how one approaches LLX, what are its scope and versatility, and what constrains its application. The broadness of the scope and the versatility of LLX mean that an appreciation of its potential and an awareness of the basic aspects will indicate its applicability for each specific case. LLX is not a natural property of systems, hence it is important first to examine the possibility of devising a LLX system to achieve a desired separation, and then to proceed to the construction of the process based on this separation.

The novelty in this approach to LLX lies in the awareness that the construction of the two-liquid phase system is the prerogative of whosoever wishes to utilize LLX for a specific separation. No one such system is unique, hence a variety of selections are feasible for the same purpose; since each selection will be different, its integration into an overall LLX process will result in one of a variety of possible process schemes. However, all these schemes will have certain basic concepts in common; furthermore, comparison between schemes requires that fixed terms of reference be defined in order to validate the comparison.

The generalized approach presented here has not required concentrating on details or specifics, even though such have been utilized occasionally to exemplify a concept or to illustrate a point. Here the logic of the analysis and the logical build-up towards the separation process have been considered of paramount importance, and this has determined the format of the presentation. In the main, therefore, only concepts have been considered significant, since data and details will follow naturally once basic aspects are defined and understood. For this reason too, multiple references have not been given since in every specific case such references will have to be sought positively, as and when required.

In process development, as in any other "master-craft" the success of the final work may be judged against its simplicity and its elegance, in conjunction with achievement of goals. LLX has a natural simplicity and elegance which should not be lost when the LLX separation is integrated into a process.

Since this is not a teaching manual nor a handbook, the author's advice to the reader is first to read it right through, as one would in any other area dealing with ideas and concepts, so as to place the presentation in correct context.

Scope and Approach

This monograph is intended for R & D personnel who aim at devising and developing novel chemical processes. (Note, chemical is used here in the most general sense.)

Since liquid–liquid extraction (LLX) is a separation tool, it is clear that the processes referred to will entail a separation operation. This separation may be the crux of the process itself, or it may be a step interposed among other steps of equal or greater impact on the process. The use of LLX may be uniquely the way in which the separation can be attained, or it may be a unit operation competitive with other separation operations. The aspect of the approach that is common to all the applications, and which perhaps distinguishes this monograph, is the author's integrated view of a "process" and its development. Here, specifically, LLX is the central point which must be fully integrated into the whole, but the approach is quite general, and the philosophy applies equally to the development of any novel process.

This monograph is not a teaching manual, nor is it a review of LLX processes. It is meant to be an exposé of the integrated research approach involved in devising and designing the LLX process. In such an approach, the separate aspects which may have a bearing on the development of the process are identified from the outset, and these are then pursued in parallel, so that the whole is pictured from the beginning and becomes more clearly defined and tangible as work proceeds. This monograph aims at showing as many as possible of the separate aspects of research which must be taken together to form the basis for the development of a LLX process.

In PD work, apart from the constraints imposed by the nature of a system and its behavior, there is also the constraint of time. In LLX, as in other disciplines, it is possible to devise limiting tests which will give definitive information delineating the area, without necessitating the scanning of the whole area. This approach is invaluable when time is important and also when a broad coverage of possibilities is best likely to pay off in the long run.

The monograph is divided into three main sections. The first section encompasses the integrated approach to developing a process (here, of course, a liquid–liquid extraction process). The second section relates to modes of studying LLX systems and presents specific approaches; however, it does not cover aspects commonly found in standard texts. The third section relates to the testing of any proposed separation scheme, to attain an acceptable degree of certainty for implementation considerations.

In Part I, the scope and types of LLX separations are discussed; where these can be applied, what yardsticks can be used, etc. Generalized identification and specification of the particular separation to be achieved—separation of what, from what, to what degree? What analogous separations are known? What alternative procedures are practised for the same separation? How will the LLX separation, if practised, be interposed in the process? What constraints are imposed by contiguous operations? How will these reflect on the choice of LLX system? What are the constraints inside the system? How do these reflect on flowsheet and equipment?

In Part II, the intimate aspects of LLX are discussed. Part II is subsidiary to Part I; this means that after having identified the separation required and decided how it is likely to be attained, the study of the system begins. For this it is necessary to select the type of two-liquid system which is expected to permit the separation to be attained by the mode of transfer anticipated. Distribution coefficients, separation factors, equilibrium curves, practical equilibrium constants, contact patterns, i.e. those aspects that relate to defining the flow diagram for a specific separation, are considered here.

Part I defines the scope and approach, Part II aims to show chemical feasibility, Part III is concerned with attaining the required degree of certainty of technological feasibility. Part III, therefore, relates to the success with which the relevant aspects of technological significance have been identified and the required extent of certainty defined, and leads then to assessment of the nature and scale of testing required for the degree of certainty demanded.

I. The Liquid–Liquid Extraction (LLX) Procedure — An Integrated Approach

Definition of the Separation

This section presents the integrated approach to developing a process which, here, entails separation by liquid–liquid extraction.

Liquid–liquid extraction is a separation/transfer procedure; if it is to be applied it is necessary first to identify the separation(s) desired, the main separation as well as the secondary ones. If a component has the tendency to transfer from one liquid to another across the liquid–liquid boundary by whatever mechanism, then separation of the liquid phases will constitute also a separation of the component that has transferred.

If A is to be separated from B, one must evaluate, conceptually, whether A should be transferred away from B or vice versa. Secondly one must define the differences in properties between A and B which can be exploited in order to achieve the desired separation. Clearly the more A differs from B the more likely will it be that a high degree of separation can be attained. Next, it is necessary to identify the property which is to be exploited for attaining the separation(s). Thus if an acid is to be separated from other acids, one needs to have some scale of comparison of the acids, so as to see whether separation can be anticipated directly from this scale, e.g. strong acids *vs* weak acids, or the tendency to complexation, hydrogen-bonding, etc.

Once one knows what is to be separated from what, and how one hopes to promote the separation, one can look at solvent selection. The characteristics of the solvent, which must interlock with the selected characteristics of the solute being separated, need to be identified. Once the characteristics of the solvent have been specified, it is possible to decide what the components of the solvent will be. Thus there are certain degrees of freedom, according to the type and number of solvent components, and the extent of interaction between them. It is possible to distinguish between the active extractant (reagent, complexing agent), the diluent (with various degrees of interaction and activity) and the modifier. The extractant is the central component and to a considerable extent determines the specificity or selectivity of the solvent. It also opens up fields for synthesis. Modification of specific reagents can increase their specificity in desired directions.

Aspects relevant to system selection are presented in a generalized form in Table 1.

According to the Oxford English Dictionary, a "process" is "a series of actions or operations used in making or manufacturing or achieving something". A LLX process is therefore a series of actions or operations which includes the LLX operation. LLX has been defined as "the process of transferring a solute from one liquid phase to another immiscible or partially miscible liquid in contact with the first". This definition implies that in LLX a series of actions or operations are used to achieve the transfer of a solute from one liquid to another. The overall liquid–liquid extraction process therefore entails probably more than one transfer operation each aimed at achieving something; taken together with other non-LLX steps the overall aim of the overall process will be achieved.

The liquid–liquid extraction operation which entails transfer of at least one component from one phase to another also implies separation of the component(s) by separating the phases along with transfer.

In order to approach the design of the process it is necessary to define what must be separated from what and for what purpose. So we may have "differential purposes", and the "overall integrated purpose" which then defines the process as a whole. The overall LLX process, which may entail a series of differential LLX steps, may itself form a part of a more encompassing process, entailing also non-LLX steps, which, taken together, achieves the ultimate aim.

Now once we know what separation we "want", we have still to identify the separation steps by means of which we shall achieve the overall separation desired, and then to identify the mode of achieving each step.

When looking at solvents from the point of view of LLX, i.e. from the point of view of reversibility of transfer and separation, one may choose selectivity or specificity as a criterion.

Specific reagents can teach by analogy in relation to procedures and interactions for general cases, but no more than this, since by definition this application is particular. The general case, therefore, must be that of selective reagents where selectivity follows some order, as a function of parameters that can be selected and controlled in each particular case.

There is conceptually a difference between LLX processes and LLX technology. For the process, where the desired operation is defined as far as possible, specificity is highly desirable; in other words, one wants to attain the ideal as nearly as possible, and there is little interest in what else could be achieved. In the technology it is important that the learning be as extensive as possible; in other words, it is preferable to have a selective reagent with

TABLE 1. System selection.

Typical stages in system selection	Specify separation desired Identify transfer required Select possible mode of transfer Choose suitable type of second phase Test "chemical" feasibility
System considerations and constraints	"Chemical" Technological Economic
System types	Both phases organic Both phases aqueous Aqueous/organic Considerations: 　　　　　　　　　　　Examples Aromatic/aliphatic separation Enzyme recovery biphase aqueous systems Water distribution Solute distribution Range of two phase zone Parameter effects
Distribution modes	Selective Specific Out of feed phase Into and out of accepting, selected phase
Typical stages in phase selection	Characterization of solute(s) 　Solubilities 　Separations for {identification, isolation, analysis} 　Valid analogies
Evaluate proposed application	Consider current separations 　procedure type 　attributes 　detriments LXX should: overcome detriments 　　　　　　　not negate attributes 　　　　　　　bring new attributes 　　　　　　　not bring detriments

controllable selectivity which can be widely applied for a variety of desirable separations. Both approaches require definition of the solute characteristics to be exploited for the separation and the matching of the extractant accordingly. However, for the process the desired separation is definitive, while for the technology only the type of separation can be definitive.

It is with intent that the operation being discussed has been termed "liquid–liquid extraction" and not "solvent extraction" for we are discussing the transfer of a component(s) from one liquid to another in order to attain a separation. The two liquids may differ greatly in physico-chemical properties, or may be extremely similar, but still different enough to provide two separate phases, i.e. not to be homogeneously miscible in the region of interest. Once the separations to be attained have been specified and the reasons stated, it is possible to look at solvent choice. The solvent choice is firstly limited by the characteristics of the components being separated, and by the fact that separation implies also recovery and/or rejection. As mentioned, it is necessary to identify the property(ies) to be exploited for the separations, then to match the counterpart solvent by defining characteristics, and thus proceed to the choice between selective and specific extractants.

Partition or distribution coefficients describe the transfer of particular components. Separation factors, which are ratios of distribution coefficients, describe the separation tendency between two components. This alone cannot, however, describe the quality of the separation, since the ratio of quantities must enter into the picture too. Thus if undesirable components have high concentration but low distribution, while the desired component has lower concentration but higher distribution, the low fraction of the undesirable component that distributes, may still cause undesirable contamination of the desired product. This is essentially the reason for considering transfer in two directions, i.e. the separation factor from phase 1 to phase 2, and the separation factor back from phase 2 to phase 3. If the equilibrium curves for the two components do not have the same shape, e.g. if the distribution coefficient in one case is a function of concentration while the other is independent of concentration, the separation factor will vary, being dependent on concentration.

The effect of a contaminant (B) on the quality of a desired product (A) in a LLX system will depend on the relative types of distribution of the two, and their interactions. A selection of possibilities is presented graphically:

(i) In Fig. 1, the distribution coefficient of (A) is considerably higher than (B) and (C) and there are no active interactions; however, the actual concentration of (B) is higher than (A), while (C) is lower than (A); the shapes of the distribution curves as a function of concentration are similar. The Figure shows the effect in the forward extraction step, in a solvating system.

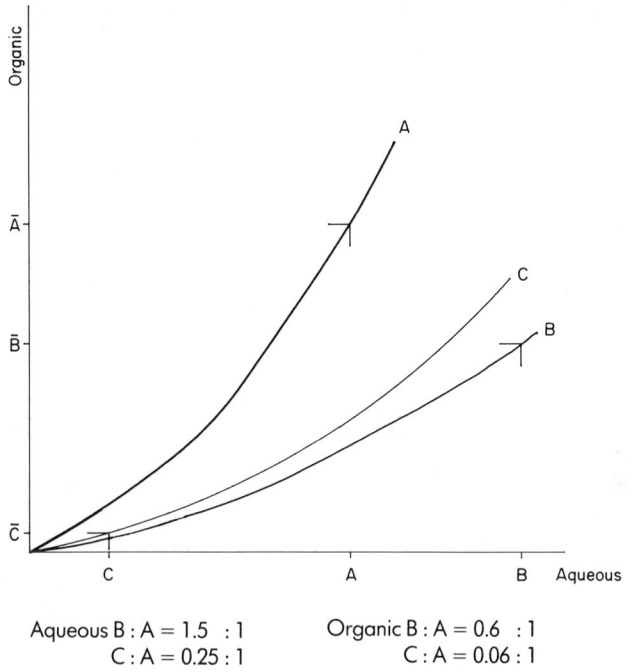

Fig. 1 Effect of relative concentrations, distributions and interactions on separations—distribution i

Ratio of (B) to (A) in aqueous feed = 1.5:1
Ratio of (B) to (A) in organic extract = 0.6:1
Ratio of (C) to (A) in aqueous feed = 0.25:1
Ratio of (C) to (A) in organic extraction = 0.06:1
(ii) In Fig. 2, the distribution of (A) again is considerably higher than (B) and (C), but distribution of (A) is strongly concentration dependent, while (B) and (C) are not.
Ratio of (B) to (A) in organic extract = 0.15:1
Ratio of (C) to (A) in organic extract = 0.06:1

The interesting aspect here is that at low concentration of contaminant the shape of its distribution curve is of little significance, whereas at high concentration it is extremely relevant.

In both cases a back wash will improve the quality of (A) considerably. Practical cases can be examined in a similar manner so as to anticipate effects and to propose procedures.

Solvent extraction is an old concept, fully integrated into organic chemistry since the earliest days. The basis for this type of solvent extraction is

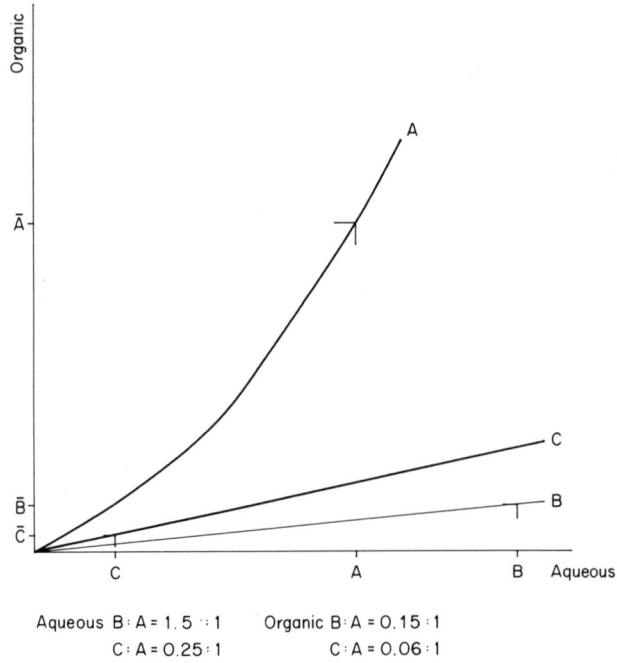

Fig. 2 Effect of relative concentrations, distributions and interactions on separations—distribution ii

preferential solubility, and its aim is recovery and/or purification. The aspect that distinguishes the concept of liquid–liquid extraction from the solvent extraction concept is that LLX relates to separation, and the two liquids are therefore intimate parts of the separation system. Strategy plays an important part in LLX processes, i.e. the careful planning towards a goal; so indeed does tactics—the skill in employing available means to accomplish an end. Goals may be many and varied, some attainable reasonably easily, others remain goals for the future. Skills are more directly definable, and these apply in various combinations, whatever the goals.

An interesting point to remember is that we are talking about "aided" distributions between two liquid phases, i.e. not just a ratio of solubilities of the molecule itself between two phases. Also the organic adage that "like dissolves like"—so useful in synthetic chemistry for isolation and separation—does not apply directly, certainly not in a simplistic manner. The reason for this is that the "system" can be compounded synthetically; all that is required is that two liquid phases should exist at the conditions of liquid–liquid separation and that the solute(s) should be distributed between these two phases in such a way as to lead towards the goal, whatever this may be.

LLX has been applied in recent years for remarkably different purposes ranging from separation of aromatics from aliphatics, through metal ion separation and recovery, to water removal for purification or concentration, to reactions, to upgrading by removing the major component or by removing the contaminants, to enzyme separations, etc. Examples of cases known to be applied on commercial scale are presented in Table 2, while Table 3 summarizes the general areas of such applications and the specific purpose of the LLX step.

TABLE 2. Examples: LLX commercial applications.

Copper
Nickel/Zinc
Rare earths
Platinum metals
Uranium
Boron
Aromatics from aliphatics
Citric acid
Phosphoric acid
Potassium nitrate
Caprolactam
Mixed aliphatic acids
e.g. formic, acetic, propionic
Penicillin
Pharmaceuticals

Rarely will the LLX operation be the whole process. Usually it will be employed for a specified separation, within an overall process entailing other steps and unit operations.

Typical cases of insertions of LLX into general processing flowschemes are shown in Figs. 3.1–3.4. The two broad areas represented in Figs. 3.1 and 3.2 show that the insertion of LLX into general processing flowsheets may follow a fairly general pattern, provided the solvent remains essentially in cyclic operation within the LLX sub-system. This pattern entails the preparation of an aqueous liquid phase containing the desired values, which then constitutes the feed to the LLX system whereby the values are transferred to another aqueous phase as product for further processing. The case presented in Fig. 3.3 differs from the above, in that this is a LLX process; Fig. 3.4 shows still another type of operation since here the two-phase system is generated *in situ* from the feed stream.

TABLE 3. LLX—process applications.

Organic
  Petroleum and petrochemicals processing
    Aromatics–aliphatics separation
    Lube oil extraction
    Extraction of caprolactam
    Acetic acid
    Xylene extraction
  Pharmaceutical manufacturing processes
    Penicillin
    Other antibiotics
    Non-antibiotics
  Food industry
    Extraction of lipids
    Decaffeination
    Extraction of flavors and aromas
  Miscellaneous organic processes for chemicals from coal and isomer separations
    SX of coal
    Phenols from coal tars and liquors
    Dissociation extraction
  Extraction reaction processes
    Aromatic nitration
    Aromatic sulfonation
    Alkylation reactions
    Hydrolysis of fats
  Industrial effluent treatment (non-metals)
    Phenol recovery

Inorganic
  Metals
    Copper
    Nickel and cobalt
    Tungsten and molybdenum
    Chromium and vanadium
    Cadmium and zinc
    Rare earths and thorium
    Precious metals
    Uranium
      Uranium and plutonium from irradiated nuclear fuel
      Uranium purification
    Zirconium–hafnium
    Niobium–tantalum
  Miscellaneous
    Phosphoric acid—production, purification
    Metathetic salt–acid reactions
      Potassium nitrate
      Other alkali salts
    Separation and recovery of halide salts from brines
    Anion exchange
    Water transfer
    Boric acid recovery from aqueous solution
    Hydrogen peroxide

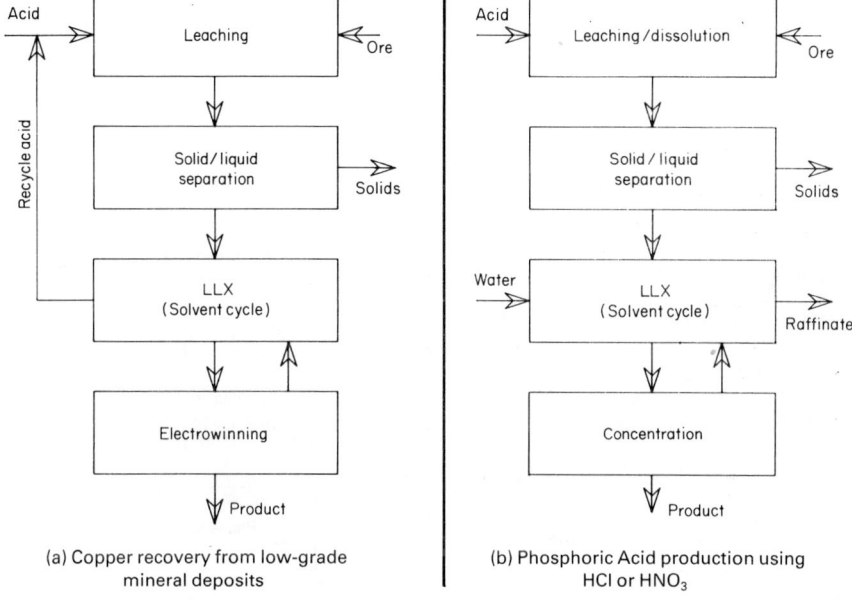

Fig. 3.1 Mineral processing—recovery of values

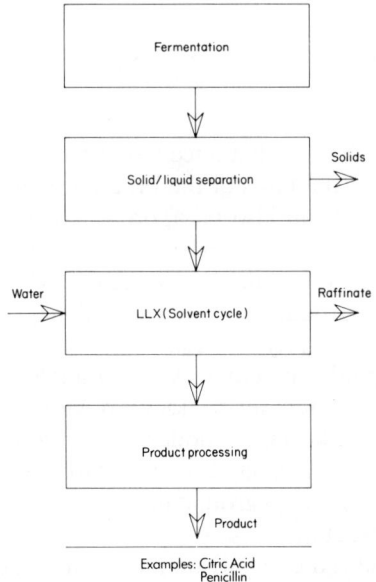

Fig. 3.2 Biotechnology—recovery of values

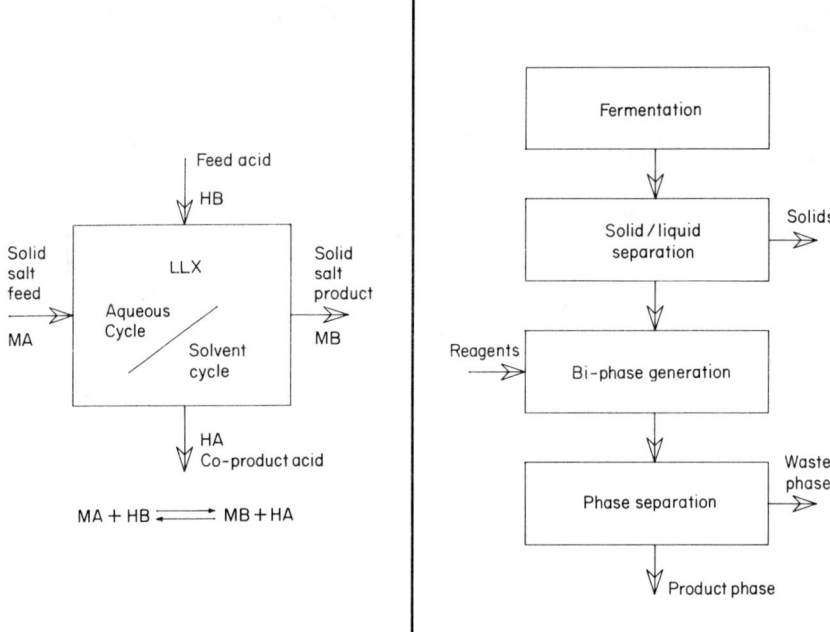

Fig. 3.3 LLX process: salt-acid metathetic reactions

Fig. 3.4 *In-situ* bi-phase generation

An important point here is that once a separation is designated it can be approached *a priori* on the basis of the broad accumulated background of LLX possibilities, but it can also be approached by analogy, drawing on other fields obviously or not obviously related. Thus, what can be done by ion exchange on resins, for example, can be done by liquid ion exchange. The latter, of course, introduces an additional dimension namely solubility, it also eliminates very largely the constraints of diffusion, which are so important in solid–liquid contacting. Another analogy would be "preparative ion pair extraction"—here the analogy is obvious, but it is not so obvious when one uses instead the more modern designation of "phase transfer catalysis"—yet in fact the analogy is considerable, not only as regards the transfer agents but also for background in order of transfer, displacement of ions by others in series, etc.

In Table 4 a generalized comparison has been made between liquid and resin ion exchange.

TABLE 4. Comparison of liquid ion exchange (LLX) with resin ion exchange (RIE).

LLX	RIE
Normal operation: steady state; counter-current	Normal operation: cyclic multiple batch
Rapid mass transfer	Diffusion rates control mass transfer
Equilibrium and working level concentrations determine phase ratios	Cycle time and break through determine resin quantity
Scrub and back wash step inherent, if desired	Displacement rinse required, causing dilution
Possibility of secondary transfer modes	Usually no solubility or secondary transfer modes, but swelling must be considered
Ease of recycle, inter-stage transfer, etc.	Recycle may require multiple columns
Mutual stream contamination by solubility and/or entrainment	No resin contamination of liquid phase, but resin fouling by feed components possible

The problem in making full use of analogy is that rarely will the persons involved have equal expertise in both fields, hence quite likely the analogy will not be sought. Also the "jargon" of one field may be quite different, thus making it difficult to spot similarities. Some inter-disciplinary transfer analogies are given in Table 5, to make the matter more explicit.

TABLE 5. Inter-disciplinary transfer analogies.

Example: Chemical processing to biotechnology
    Electrodialysis ⎱ from desalination to separation/concentration of
    Reverse osmosis ⎰ fermentation products
    Paper chromatography from analysis for cations, to synthetic adsorbents, reverse phase chromatography, high-performance liquid chromatography (HPLC) etc. for amino acids, proteins, enzymes
Example: Mineral processing to biotechnology
    Selective aggregation, flocculation, flotation, fluidization
Example: Chemical engineering to biotechnology
    Scale-up
    Continuous steady state processing and operation
    Process design and control

Liquid–Liquid Extraction. Its Nature and Scope

LLX is no longer what it started out as, when it was a unit operation of very little applicability, warranting only the smallest fraction of the attention devoted to other unit operations of more major interest such as distillation and heat transfer. Nevertheless, separation and recovery from solution was a very basic requirement in chemical processing—in all its variations and branches—so the potential for applying LLX was in fact extremely broad. The restrictiveness of the thinking in relation to two liquid phases was what confined the scope of application of LLX. This thinking assumed that two liquids of extreme difference in characteristics were required for satisfying the primary condition that two liquid phases exist, and this notwithstanding the equally well accepted fact that in the neighbourhood of a plait point, the two liquid phases will have very similar compositions, i.e. the extreme difference in characteristics certainly does not apply as a necessary condition for the co-existence of two liquid phases. The real point is that the "virgin" condition of two liquids which are to constitute a part of a LLX step is in a sense of limited relevance, rather the significant point is the composition of the two phases after contact, since there will be some degree of mutual miscibility even in absence of the solute which is to transfer between the phases, and this changes the basic characteristics of the virgin materials.

Essentially, therefore, it is the components of a system and their mutual interactions that determine the range of multiple phases attainable. In the reverse, therefore, *a priori* selection of these components based on an understanding of their interactions will ensure the presence of two liquid phases; hence, in principle, will permit the constitution of LLX systems for the purposes in view.

Phase theory and the thermodynamics of heterogeneous equilibrium have been addressed; also considerations of solvation, coordination, complexation all are positive steps in the direction of eventually arriving at practical LLX systems. The more that is understood about demixing in multi-component systems, about distributions in multi-component systems, and about the promotion of distributions, the more will the selection of practical systems for specific purposes be facilitated.

In one of the best studied examples for applying LLX, namely the extraction of Cu^{2+} from dilute aqueous solution, no-one would dream of describing the system as a "water/hydrocarbon" two-phase system, since in fact it is the Cu^{2+}, SO_4^{2-}, H^+, $\overline{R}^-$, and the equation

$$CuSO_4 + \overline{2HR} \rightleftharpoons \overline{CuR_2} + H_2SO_4$$

which describes the system, i.e. the tendency for the equilibrium to move to the right by virtue of the possibility of the $\overline{CuR_2}$ thus formed dissolving in a suitably selected second phase, provided that the H_2SO_4 level in the aqueous phase is low enough not to reverse the reaction; hence the second phase is the $\overline{HR}$/diluent mixture, certainly not the diluent as such, nor $\overline{HR}$ alone. If for any reason a modifier is added to the second phase, this must be regarded as an intimate part of the system, and the distribution of Cu^{2+}, or in practical terms its extraction efficiency, must be related to the total second phase.

Recognition of the wide scope of LLX for separations is an exciting consideration, but it may also seem to be a formidable challenge—both of these aspects derive from the same characteristics, namely the freedom attached to a learned selection and the broadness of the choice. There is no point in assuming or hoping that the system one needs for a particular separation will have been described; even though a fair number of extractants for specific purposes have been studied, this is not even the smallest part of possible systems, so this is no way to look for a workable LLX system for one's particular case or problem, expecially if it does not fall into one of the limited classes of separations that have been the scene of major activity thus far in LLX.

What one needs to do is to look for a clue to selection. The clue may appear far fetched initially, but once one can explain even an esoteric case, one can immediately make practical selections by analogy.

Usually the clue to possible extraction routes will be found first by an assessment of the chemical nature of the separation being sought, or by examining current modes of attaining this separation in any context—analytical, preparative—or by using an analogous case as guideline.

Take as example the separation of iron from an acidic medium—the chemistry of Fe shows that Fe^{3+} is complexed in the anion in acidic media far more readily than Al^{3+} for example, and more readily than Fe^{2+}, hence acid extraction will separate $FeCl_3$ from $AlCl_3$ or $FeCl_2$ since the species $HFeCl_4$ will be extracted. By analogy, then, Cu^+ would be separated from say Fe^{2+} by a similar acid extraction in a reducing medium, since CuCl will be extracted as $HCuCl_2$ preferentially over $FeCl_2$.

Another example would be extraction of an amino sugar which is currently being recovered by cation exchange using a strong resin exchanger. This immediately suggests that a liquid ion exchanger could probably be used similarly for LLX, provided the reagent selected is sufficiently strong; this leads to considering a sulfonic acid as ion exchanger since the order of acid strength is known to be sulfonic > phosphoric > carboxylic, which are the commonly available liquid ion exchangers. This also leads to the question whether a phosphonic or phosphinic acid would be of suitable strength, or an α-substituted halocarboxylic acid, which is a stronger acid than the unsubstituted parent acid.

Let us take yet another example—penicillin has classically been recovered by extraction with solvating solvents being eventually released by neutralization. Analogy with other cases of extraction of acids says there are two approaches to be considered; using solvating solvents or formation of ion pairs. Ion pairing is probably more favorable for extraction but more difficult for stripping; however, neutralization obviates the stripping problem. For a case like penicillin therefore, if all other specific constraints can be met, ion pairing should be a favorable procedure.

A different type of approach to finding a clue comes from the work that was done on water extraction from brines using amines; the marked effect of temperature on the mutual miscibility in the amine–water system was utilized as the means of water release. This leads to considering other solvents with similar characteristics as extractants, particularly for cases where a hydrophilic extraction environment must be maintained. Trade literature is a very good entry point for someone seeking solvents by analogy to amine–water systems.

Let us take an entirely different approach. Suppose someone has published material showing that a particular solvent is good for a particular extraction without specifying anything about the nature of the interaction between extractant and solute. One way of attempting to arrive at other likely candidates for a similar procedure would be to see where the cited solvent is being used specifically and what other known solvents behave similarly, and then to try to generalize the characteristics of these solvents. Thus all solvents proposed for separating aromatics from aliphatics could be considered as belonging to one class even though this classification has not been specifically defined. If one of the aromatic extractants is a potential solvent for something entirely different, then logically others from the class should be considered too.

In this way one goes by analogy from a known case to an unknown one, expanding the horizons for LLX.

Once a system has been selected, the mode of studying it for arriving at guidelines for design is no different from any other case. In other words, the R & D and PD procedures, equipment testing and selection, will follow well-charted routes; it is the selection of the mode of transfer and the second phase composition to go with it that are the challenge in every new case.

There is another interesting aspect—reagent synthesis and modification has played a major role in the success of the Cu^{2+} case history, but this is hardly so in other LLX success stories. There, usually an understanding of the chemical interaction required or desired and selection of likely materials from among well established available compounds has been the line followed. From this one learns that existing compendia of available materials will satisfy the first requirements of any novel application; only later, for optimization, or for satisfying specific targets or the constraints of a special case of a well established, well documented type like Cu^{2+} extraction, is it really necessary to go into reagent modification and synthesis.

In this respect, LLX differs from adsorption and from solid resin ion exchange where considerable specific synthetic expertise is required to develop adsorbents and ion exchangers. The selection of components of the second phase of a LLX system requires a general understanding of chemical interactions, but need not entail synthetic chemistry in the first instance.

Solvent Components; Extractant (Reagent), Diluent, Modifier

In the simplest cases, the solvent may consist of only one specified component, although in fact in a steady state cyclic process it is highly unlikely that the solvent will ever return to the initial composition at time zero. Rather, perhaps, one can say that make-up will entail addition of one material only. Again, clearly this need not be a pure compound, but its composition should be consistent.

The single solvent offers limited room for manipulating the system since it alone must meet all process and operational requirements. In other words, it must satisfy all aspects that will lead to an overall viable system. These aspects include selectivity, capacity, solubility, mass transfer, phase separation, costs, among others.

An extension of this type of system is the mixed solvent system, where two types of solvating solvents are used in order to change the mutual miscibility characteristics. An example is given in Table 6.

TABLE 6. Phosphoric acid/ether plus modifiers.

$H_2O/S_1/S_2/Acid/Ether$ + Solvating Solvents
$S_1 = nBU_2O$ $\quad\quad S_2$ = Modifier
Acid = H_3PO_4
Range of K_D = 0.01–0.5

S_2	Aq. phase H_3PO_4 wt. %	$K_D \dfrac{\text{org.}}{\text{aq.}}$
None	74	0.01
	81	0.45
7% Propan-2-ol	47	0.01
	73	0.3
10% Butan-1-ol	57	0.02
	75	0.4
15% Hexan-1-ol	58	0.04
	77	0.4
13% n-Octanol	57	0.02
	77	0.3

The main aspect here is that the primary solvent determines the nature of the distribution, for example, in this case the distribution of acid is very strongly concentration-dependent. The second point is that modifiers of the same type behave essentially identically, although the quantity required may initially be dependent on molecular weight of the modifier, here the —CH— chain length, but this becomes less significant as the chain length increases.

The next simplest system is composed of an extractant and a diluent. The extractant may be an ion exchanger or a specific complexing agent; the diluent is regarded as inert, and is assumed to influence only the physical characteristics of the system, especially viscosity, but it helps also to solubilize the extractant–solute complex. Since diluents are never wholly inert, active diluents may in fact be selected by intention. Addition of secondary diluents or modifiers leads to less simple solvent systems, where interactions within the solvent phase can provide a gradation of properties which may make finer separation possible. These interactions may affect the relative acidity or basicity of systems, the tendency to form a second solvent phase, solvation of the solute, etc.

Solvent systems composed of at least two components offer more possibility in selection than does a single component. Thus in a two-component system it is possible to modify or change the extractant for optimization; however, since the diluent essentially determines the physical properties of the compounded solvent, a change in the extractant does not mean that global properties such as viscosity, specific gravity, interfacial tension need change. This is important for operation since it ensures continuity in development. Positive selection of a multicomponent interactive system has come to be more widely accepted only recently, even though the phenomenon of synergism or antagonism was recognized since the earliest work on uranium extraction. Also the effect of the modifier, which is added to prevent formation of a third liquid phase, has been revised since in fact it may be a co-solvent, interacting with the extractant directly and thus changing its characteristics.

In recent years it has been generally accepted that no "diluent" for any extractant is ever "inert" and increasing emphasis is being placed on diluent effects, particularly the effect of diluent polarity and polarizability. The use of modifiers, especially alcohols, to prevent third phase formation in practical extraction systems is well known, but very little, if any, attention has been paid to other modifiers even for the same purpose of "third-phase suppression", and the distribution coefficient as a function of diluent and modifier has hardly received attention.

Modification or adaptation of the solvent system will usually be motivated by process or economic advantage. Process advantages are mainly improvements in separation and selectivity. Economic advantage will relate to a smaller quantity of solvent in cycle per unit separated, less costly solvent, reduced losses, better phase separation, faster kinetics, fewer hazards, lower investment, etc.

A somewhat different type of two-phase system is what has been called "aqueous polymer two-phase systems", used for partitioning proteins and enzymes. These are compounded systems, both phases being very high in water content. Conceptually, two-phase systems formed by water and, for example, glycol ethers are similar in the sense that the water distributes but the glycol ether prefers one phase. So too with the polymers, each prefers one of the phases, whereas water distributes almost equally as a function of temperature. However, with respect to the partitioning component the two phases are different, even though from the water content point of view they are similar.

Extractant Characteristics; Types, Interactions, Specificity, Selectivity

In principle, choice of extractants would seem to be limitless in scope. In fact, of course, this is not so, for reasons of availability, safety, costs, etc. Nevertheless, if we accept the definition of an extractant as an organic compound capable of acting as a promoter of transfer in a two-liquid phase system, and as having the characteristic of selectivity in the specific case under consideration, then indeed the examples are legion. Thus a text on coordination compounds will open up a very broad choice; mutual solubility tables will show a catholic collection of compounds. Similarly, if one were to list all the reagents studied in "solvent extraction" one would have a fair number to consider. This cannot be the mode of approach, however, since it leads nowhere, or at best only helps one to wander in the forest without showing where the paths are.

An approach which looks, instead, at extractant characteristics is therefore more legitimate, provided it is combined with modes of selection and interactions. Thus extractants may be inert, or may be graded according to polarity, hydrogen-bonding ability, donor number, etc. or they may be acidic or basic, or chelating or ionic. These can then participate in selective extraction by compound formation, solvation or ion pair formation as the case may be, according to the environment. Specific reagents are common in analytical chemistry, but one should not forget that the conditions are usually specified very clearly for such specific reagents. Also they may be specific for a family of compounds, which will therefore not necessarily satisfy the separation desired, when the separation required is between similar materials. Once conditions are defined within narrow limits, specificity and selectivity may become almost indistinguishable; in other words, under defined conditions the reactant will promote transfer essentially only of the component desired, even though under different conditions this would not be so.

Solvent Selection

Practical industrial processes, developed and implemented during the last 20 years, show that not only can LLX be applied for a wide variety of purposes, but that the second liquid phase can be selected, compounded, tailored to fit the case. This is a far cry from the classic picture of solvent extraction which consisted of water, a simple organic liquid and a solute, and which was presented simply as a ternary diagram. Still even now the ternary diagram is a very elegant way to represent a two-liquid phase system, but the apices must be defined specifically in order to make the concept of "three" fit, i.e. liquid phase I, liquid phase II and the transferring solute(s). However, tailoring second phase to fit is not as simple as it may sound, since there are various and varying constraints on choice according to the circumstance and the application.

Probably the most important aspect of choice is the defining property of the component being selected. Since LLX implies distribution or partition of a substance or substances between the two phases, and furthermore the purpose is to arrive at favorable distribution, where "favorable" is to be interpreted within the context of the case in point, the defining property will vary from case to case. If we take "organic" as one end of the line and "aqueous" as the other, then both or either of the phases may be wholly or mainly organic, or essentially mainly aqueous, or range between the two. The number of combinations of materials which are not homogeneously miscible in all proportions, and which will result in two liquid phases is large; furthermore, the range over which a particular combination will give a two-phase system and not a homogeneous single liquid phase may vary from very narrow to very broad, it may be symmetrical relative to an imposed parameter which causes demixing, or skewed in one direction or the other. The distributing component too may have a marked influence on this demixing zone, in either direction. Also physical properties such as density, viscosity, interfacial tension, surface tension, may change considerably as function of composition, temperature and probably pressure, and all these may have direct or indirect interaction with the transfer required and the mode by which it is attained.

Furthermore, in application, the characteristics of the two-phase system, its sensitivity to demixing and the range it covers as regards physical properties may have a real bearing on the design of the process flowsheet and on the selection of equipment. At the present technology level of LLX in chemical or hydrometallurgical processing, the so-called solvent phase will usually consist of the active reagent, often dissolved in a diluent with a modifier added. This usually means that this phase is described as the organic phase, irrespective of the level of water dissolving in it. In chemical and hydrometallurgical processing the given feed phase will always be considered as aqueous phase. Recently, biotechnology has brought a different concept of two-phase systems, as of course has always been the case in LLX in petrochemical or organic chemical processing, where both phases may be wholly organic. In biotechnology, both phases may be essentially aqueous.

In chemical or hydrometallurgical processing, the defining property of the main or active component of liquid phase II can be described in various ways. This main component is assumed to be the one which always interacts positively with the material distributing, while diluent and modifier will interact with this main component itself, but may also interact with the distributing material, directly or indirectly. Still a further interaction will be that between the solvent of liquid phase I (the given phase) and phase II (the selected phase).

The type of property defined may be ion pair formation, ion exchange, basicity, acidity, polarity, donor properties, dielectric constants, chelation/complexation, etc., mutual miscibilities, specific gravities, viscosities; the dependency of particular properties on external or imposed parameters such as the temperature, pressure, acidity levels will be relevant. Also important is the extent to which the interaction of the main component with the distributing component affects any other property. Thus polarity, mutual miscibility, specific gravity, viscosity may all be strongly affected by the level of the distributing component in the second liquid phase.

Returning to the main component of phase II: its defining property is assumed to be that which promotes the transfer desired. Thus if an acid is to be extracted one needs a basic transfer agent or extractant and vice versa. This means that a good understanding of the partitioning material is necessary in order to be able to select the correct counterpart for promoting transfer. At the same time, not less important is a good understanding of the properties and behavior of accompanying materials which should also transfer or should not, as the case may be. If behavior of accompanying materials is similar to the partitioning material, it may be desirable that the counterpart extractant be specific or at least selective in its interactions, so as not to promote transfer of unwanted materials.

Up to this point interactions within the LLX system itself have been considered; however, in selecting the second liquid phase, its interaction, effect and repercussion on aspects extraneous to the LLX system must be considered too. In general these will derive first of all from the level of solubility of phase II components in phase I (the given feed) and in phase III (the produced LLX product phase). This will be relevant whether residual phase I (raffinate) is recycled, processed further, or discarded. Similarly, phase III will be processed according to its nature and the scheme of things, hence here too there may be considerable importance as to even trace quantities of phase II components that transfer or are carried over into phase III. There may be safety or environmental codes in various industries which do or do not permit certain substance types to be used—this aspect is paramount in the food and pharmaceutical industries.

Tailoring phase II to suit the case brings additional problems in its wake. As long as the solvent phase in LLX was a simple organic liquid or a mixture of similar liquids of specified boiling range, it was easy to control process solvent quality and composition by evaporation/distillation procedures. For compounded solvent such a step may be out of the question, but solvent quality and composition is possibly even more important in complex, multicomponent systems. If evaporation can no longer be applied one must usualy have recourse to chemical clean-up procedures, possibly to crystallization or precipitation, or adsorption. The more complex the system, the less likely will it be that it will return to the composition at zero time. Hence properties of the virgin mixture may not be representative of the phase II in operation. This aspect of change may be positive or negative, but in any event cannot be ignored.

System Properties and Their Significance

For liquid–liquid extraction to function it is necessary to have contact between the two liquids, so that transfer of material across the liquid–liquid boundary can take place. Two aspects control the transfer, one relates to equilibrium, as expressed by some equilibrium constant, e.g. the distribution coefficient, the other relates to the kinetics of transfer and is controlled by the concentration gradient as driving force, and the area of contact between the two liquid phases.

The system properties therefore have significance, to the degree that they reflect on transfer. The property of the system that controls the area of contact is the interfacial tension—it has been said that this is "the most significant physical property for two-phase liquid systems".

Apart from transfer, the functioning of liquid–liquid extraction may depend on dispersion, coalescence and phase separation. It should be clear that this aspect constitutes an operative factor, but does not directly have an influence on mass transfer. However, to the extent that a dispersion of one phase in the other is to be used for generating contact area, the subsequent coalescence of the droplets of the dispersion and its separation into two liquid phases becomes important for continuing the operation. In the limiting case, where a solvent is permanently held in a phase which is essentially not liquid, the whole aspect of drop coalescence and phase separation falls away. Similarly when two liquids flow as thin films in contact without dispersion, coalescence has no place in the system.

When one regards the second liquid phase in a LLX system as a transfer medium, i.e. material goes in and out, then at first glance the ultimate would be a thin layer of this medium which picks up and delivers. This is what membranes, whether solid or emulsion type, are ultimately expected to do.

When the transfer is coupled, as, for example, in cation extraction using hydrogen ion concentration as the driving force, there is conceptually no problem in visualizing a gradient, e.g. of H^+, to promote transfer in the correct direction across a membrane. A complication here would be H^+ leakage in the reverse direction, tending to negate the transfer.

If a temperature gradient must be maintained for controlling the direction of transfer, this becomes a more difficult problem when membranes are to be considered, since the barrier to reversed heat transfer will be minimal. If material transfer is more rapid than heat transfer, it may be possible to keep a temperature gradient across the membrane even though there is real heat loss across the membrane. Presumably, also, in the case of membranes, from the kinetic point of view only mass transfer rates need be considered, since coalescence rate is no longer a factor. It is true that with emulsion membranes both dispersion and coalescence must be considered technologically, but these steps have been separated from the mass transfer itself; on the other hand, in LLX both dispersion and coalescence are an intimate part of the mass transfer process.

Some aspects of the difference between liquid emulsion membranes and LLX are given in Table 7.

The main conclusion from this cursory analysis is that LLX and solvent membrane transfer belong to the same overall transport system but represent different forks in the road. If one were now to consider as many as possible of the pros and cons of both forks, possibly one would come to an overall viable program aimed at a single transport system, optimized to attain the principal aim. The problems that eventually express themselves in the detrimental or negative aspects of the particular transfer system are essentially technological, being translated in fact into negative aspects of execution.

In coupled H^+/cation transfer conceptually the "extraction" system is controlled by the limiting compositions at each end, i.e. at the feed end and

TABLE 7. Comparison between liquid emulsion membranes and LLX.

Membrane system	LLX
Scrubbing not possible	Scrub of extract standard procedure
pH regulation difficult	Inter-stage pH control a function of equipment type
Restricted applicability of solvating extraction	No restrictions on applicability of transfer mode
Minimum inter-phase contamination	Inter-phase contamination inherent in LLX—controlled by equipment type and physico-chemical properties of system
Low volume of organic components relative to aqueous feed volumes	Finite ratio of organic to aqueous phase

at the raffinate end, in relation to the transfer agent, and by the driving force of H^+ for the "stripping system". This is so whether a bulk extractant or a fixed extractant is used. It is important also to keep in mind that the degree and concentration of recovered material will be favored by a countercurrent operation, hence in this regard, too, LLX and membrane transport must be compared.

The properties which affect coalescence most strongly are interfacial tension and viscosity. While interfacial tension has virtually no temperature coefficient, viscosity is strongly temperature dependent. Also viscosity is very much influenced by the composition of a phase, and hence may vary considerably in multi-component liquids. Mass transfer rates may be affected by whether transfer is into or out of a drop, which then in a specific case can be interpreted as relating to the dispersion type, i.e. "oil-in-water" or "water-in-oil". However, coalescence is by far the more sensitive to this factor—extreme differences in coalescence rates may result from inverting the dispersion type. Specific gravity difference is a positive factor in phase separation, quite small differences between the two phases being enough to promote operational separation.

The rate of mass transfer is dependent on the interfacial area of contact of the phases, the mass transfer coefficient which measures the specific rate of transport of mass across the interface, and the concentration difference of the transferring component which constitutes the driving force for transfer. The mechanism of mass transfer certainly plays a part in the rate of transport of mass across the interface.

Drop to drop coalescence is generally enhanced when transfer occurs from the organic drops to the aqueous continuous phase, and is retarded in the opposite direction.

When the two-phase system is generated by dissolution or admixing of some compound, then the whole question of distribution rates or mass transfer rates takes on a somewhat different aspect, as does drop formation and coalescence. A number of questions can be asked which need answering, if design approaches deriving from liquid–liquid contacting and separation are also to be applied when two-phase systems are generated *in situ*. It is of course possible to consider cycling an equilibrium phase say in a two-polymer aqueous system; here two problems arise: the one is whether such a cyclic system can be devised, the other is how mass transfer rates are likely to be influenced.

The Shape of the Equilibrium Curve and Its Implications

Equilibrium isotherms, showing the concentration of a distributing solute in one phase relative to its concentration in the other, give a practical presentation of the distribution behavior of the solute as a function of its concentration. In an ideal system, this would be a straight line relationship. However, this is rarely so over a fair span of concentrations. In solvating systems, the presentation will be as in Fig. 4(a), i.e. at increasing concentrations, but at low absolute concentrations, there is a straight line relationship; however, as concentration increases still more in one phase, the conjugate concentration in the other phase goes up much more sharply. In LLX for ternary systems of this type, "correlations of tie lines" show two distinct sets of distributions as can be seen in Fig. 4(b). This is typical, for example, of inorganic acids distributing between an aqueous phase and alcohols or ketones.

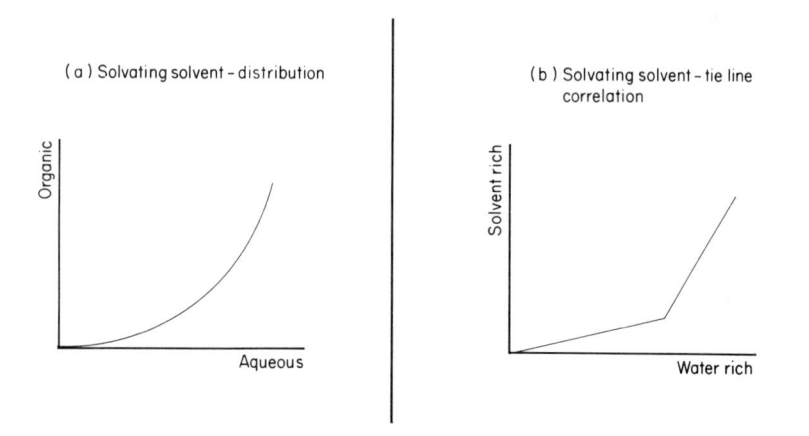

Fig. 4 Typical distribution curves; (a) Solvating solvent–distribution; (b) Solvating solvent–tie line correlation

When the mechanism of extraction is ion-pair formation, or when a neutral species of defined stoichiometry is extracted, the distribution isotherm will have a different shape, as in Fig. 4(c); the concentration which can be attained in the solvent phase will be a function of the reagent concentration and the stoichiometry.

In ion exchange, or metathetic reactions, the equilibrium is a function of the two ion concentrations, hence this type of distribution curve cannot be used conveniently; instead the concentration of the extracted ion will be plotted against the exchanged ion in the aqueous phase. Hence, the way in which cation extraction is usually presented is the percentage of total cation which reports to solvent phase as a function of pH of the aqueous phase, as shown in Fig. 4(d).

This type of curve shows the range of pH in which some desired measure of extraction can be attained. However, the material balance or the absolute phase ratio is tacitly inherent in this presentation. This is not so in ion-pair formation or when a neutral species of defined stoichiometry is extracted; what the equilibrium curve shows in the latter cases is the ratio of bound to free reagent. At some limiting concentration in the aqueous phase, the reagent will be essentially fully bound; after that more reagent must be used to bind more of the material to be extracted. The molecular weight of the reagent and its specific gravity gives the maximum concentration of extracted material attainable in the solvent phase. In practice this is usually not a working proposition, because of viscosity of the complexed reagent species and the melting points of both it and the free reagent. Usually therefore a diluent must be used which is a solvent for the free reagent and for the

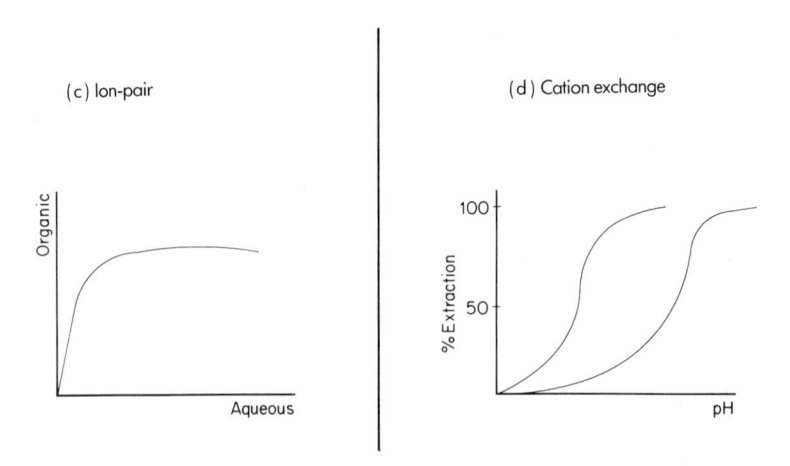

Fig. 4 *contd.* (c) Ion-pair; (d) Cation exchange

complex. Often it is difficult to find a diluent which is "inert" and at the same time shows complete miscibility with the ion pair or complexed reagent. Hence, at higher bound reagent compositions an additional phase may be formed. This can be a solid or a liquid, according to the case. To prevent this, a modifier is usually added. It must be borne in mind that modifiers will not be inert, hence it is essential to know the modifier/reagent interaction for proper selection and evaluation. No diluent for any extractant is inert. Effects may differ greatly, e.g. according to polarity, polarizability. In Table 8 a limited number of cases have been listed as examples.

TABLE 8. Diluent/modifier.

Examples:	
Acid extraction	Protogenic diluents/amines
Anion exchange	Amphiprotic diluents/quaternary ammonium bases
	Amphiprotic diluents/amines
Cation exchange	Amphiprotic modifiers in cation/extractant system
Extraction of metal complexes	Separation of metal values from sulfate solutions by amine extractants in diluents of increasing polarity
Mixed solvents	Salt extraction from brines
	Water/salt ratios in extract

Returning to the shape of the equilibrium curve, this gives an immediate indication of the nature of the flow diagram as regards extraction, and permits a material balance to be calculated. If appreciable volume increase takes place in the phases due to mass transfer, it is better to express concentration on a weight basis relative to the feed solvent phase.

While the extraction curve is the one which usually receives major attention in comparing reagents for a particular task, it is, in fact, very often the stripping or back transfer that must determine the choice. The shape of the stripping curve is then the one which is truly determining.

In a solvating system, the extraction and stripping isotherms are identical; where there is little temperature effect within a feasible temperature range, the only way in which the transfer can be reversed is by sacrificing concentration or by utilizing a relatively larger number of transfer stages. In effect, this means that the operating aspects, not the equilibrium aspects, are determining. This is not so in the case where interaction takes place; for example, a common ion effect, as shown in Fig. 4(e); thus in the system $CaCl_2$–HCl–H_2O–alcohol, where extraction of HCl is favored by the presence of the

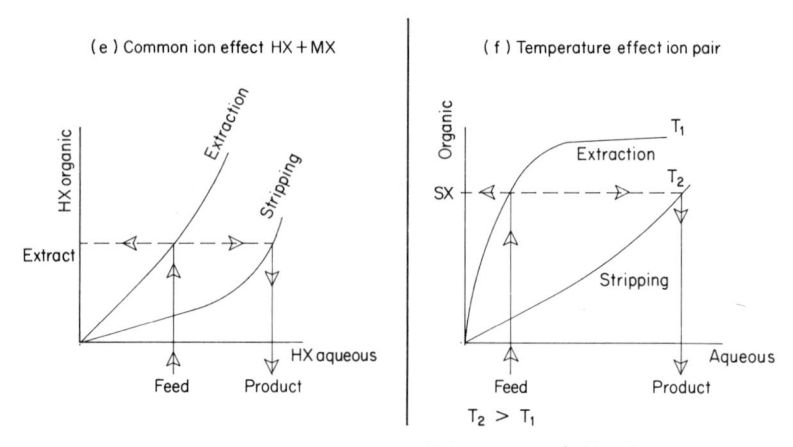

Fig. 4 *contd.* (e) Common ion effect; (f) Temperature effect ion pair

non-extractable CaCl$_2$ while stripping is favored by its absence, stripping may very well give a more concentrated acid product (though not a more concentrated chloride, of course).

In the cases where ion pairs or stoichiometric species are extracted, it is possible to eliminate completely the concentration consideration, if some other parameter controls the transfer. Thus temperature has an effect on the pK's of weaker acids and bases; hence a difference in temperature may serve to distinguish completely between extraction and stripping in some ion-pair cases, as shown in Fig. 4(f). In stoichiometric species extraction, the common ion effect may be a dominant consideration as in the case mentioned above, so that the shape of the stripping curve will resemble that of extraction in a solvating system. Hence, stripping and extraction here are essentially independent and each can be optimized for itself.

In ion exchange, the driving force for stripping will be the reversal of the ion exchange reaction. This is well established in cation extraction of copper for example, where some level of acidity reverses the transfer. In metathetic reactions, it is not possible to generalize since each case must be evaluated for itself. In all cases, however, the aim will be to attain a viable cycle, satisfying the particular purpose of the transfer.

Dilute Solutions

The application of LLX for recovery/removal of solutes from dilute solutions has a number of aspects, not all positive. So before considering the choice of solvent for a specific case, it may be as well to review these basic aspects.

LLX, since it requires direct contact between two liquids, will always entail some cross contamination, i.e. the raffinate will always take some level of the "solvent" phase with it, not always at the same composition as the solvent phase itself. The reason for this is that absolute coalescence and absolute phase separation cannot be attained in practice; also absolute non-distribution of the components of the solvent phase is unattainable. So even in cases where there is extremely efficient recovery/removal of the low concentration solute from the feed, so that this component will be perhaps orders of magnitude lower in the raffinate than in the feed, the level of overall raffinate contamination may not be less than the level in the feed itself, although the nature of the contaminants in the two cases may be entirely different.

This means that with dilute solutions there may be two types of problems, their relative significance depending on the nature of the initial low concentration solute and the purpose of, or the reason for its extraction, and depending also on the nature of the "solvent" phase selected to do the extraction, and the nature of the contamination it imparts to the raffinate. Furthermore, the definition of dilute solutions obviously varies from case to case, being defined basically within the framework of the case and its reference levels of concentration, and these can certainly differ by orders of magnitude. Thus, for example, Cu^{2+} is in low concentration in leach liquor, when it falls below 1 g/l, but toxic trace metals will generally be in the parts per million range, looking at water quality as the reference. In biotechnology we have fermentations which produce tens of percents of some specific compounds, while other cases may produce levels of active materials that are not expressed on a weight concentration basis at all, but, for example, in the case of enzymes, on an activity basis.

When dealing with low concentration feed streams, it is clear that distribution as a function of concentration is not a viable approach unless a volatile solvent is used which can then be vaporized in whole or in part to leave a concentrated solute as product. A more useful approach is to choose a solvent which shows, for example, great dependency of mutual miscibility on temperature; in other words, the homogeneous extract will be converted into a two-phase system by a temperature change, with the solute transferring back to the freshly generated phase. However, these properties, and especially a favorable range of such properties, will be more the exception than the rule, hence distribution or partition is hardly a basis for dealing with dilute solutions. An alternative approach is to use partition for extraction, selecting a very favorable extractant with a high distribution coefficient, and then to use a chemical step for release from extract. The case of Cu^{2+} recovery exemplifies this approach exactly, where a high acid concentration in the release step wholly controls the final Cu^+ concentration in the product, thus separating it entirely from any dependency on the dilution in the initial feed. This is the approach that would be used for removal of traces of toxic cations from water, namely a specific or strongly selective cation-exchanger/complexing agent, followed by acid stripping into a small concentrated acceptor phase. Equally, in certain cases, a redox system can be selected to drive the transfer back and forth.

It is not necessary to use LLX as the sole separating method; thus in biotechnology, where high value, low concentration materials are to be recovered, a combination of procedures is often the rule. Ultrafiltration, reverse osmosis, electrophoresis, electrodialysis, adsorption, ion exchange, ion pairing, any of these can be combined with or without LLX, whether this be a two-aqueous-phase partition or a more usual extraction type system.

Let us now return to the problem of cross contamination of raffinate or of product by the solvent phase itself. First consider the raffinate, which may be a waste stream or a recycle stream. The consideration of contaminants will not necessarily coincide in the two cases. In any event, the possibility of elimination of contaminants in both cases must be considered in conjunction with the initial selection of the extracting phase. Thus steam stripping, adsorption, back extraction may all be possible approaches technologically, but they may not be viable approaches economically in every case. Product contamination may have additional aspects, for example its influence on downstream operations, or on product quality, or considerations of food, drug or health restrictions in specific cases.

Although many of these aspects apply in general in LLX, the case of dilute solutions may be more stringent, because of the fact of dilution which will bring in its wake considerations which are not necessarily so relevant in the cases of concentrated feed streams.

To reiterate—the transfer agent or reactant selected should have a very high specificity or selectivity for the low concentration material to be extracted, and it should be sensitive to some parameter which can change the distribution coefficient very significantly. The diluent and modifier, if these are to be used, must cause minimum cross contamination or permit easy elimination.

Flowsheet Delineation

For LLX to be effectively used, it is necessary to know what is to be separated from what, and also the desired quality of the separated material. This means that there is both purpose and reason for the separation. Assuming that the primary liquid is the feed stream which is not open to choice, then the flexibility in LLX must lie in the choice of the second liquid, and in the way in which separations are manipulated and fitted together.

As long as one remains within the framework of LLX, all separations must be made to or from one liquid or the other. For convenience, let us talk about the "aqueous" and the "solvent phase" without restricting the "solvent phase" in any way. Let us also for convenience designate the heavier phase as the aqueous phase and the lighter phase as the solvent phase, even in cases where the organic liquid and water have appreciable mutual miscibility so that composition-wise it may be less clear that one is aqueous and the other organic. Naturally the reverse designation must apply where the organic liquid itself has a higher specific gravity than water.

In order to delineate a process, the purpose of the process should be clear; it should also be clear why one considers using liquid–liquid extraction as a central operation. In every case, there will certainly be other modes of attaining the targets, hence there must be a rational formula for deciding when to use liquid–liquid extraction, or at least a rational procedure for making comparisons with other options. One should try to keep things as simple as possible. By analogy with other processes, one should try to specify the steps which will be involved in the liquid–liquid extraction operation. Essentially this means how many times one expects to transfer to and fro between aqueous and solvent phases, each transfer contributing to separation between components. Since it is always desirable to evaluate a proposal at as early a stage as possible, it is necessary to use approximations and simplifying assumptions, so as not to overload the project with information not relevant to the first delineation.

Cost comparisons need to be made in all cases. There will be some baseline costs for each of the two systems which may rule out consideration for particular purposes; beyond this, investment and operating costs as a function of the cases need to be compared.

In any evaluation of a transfer for separation purposes the quantities and concentrations of the components to be separated must be taken into account and the degree to which separation is required.

The practical approach to developing a process assumes a specific problem. Thus when the solution contains components which must be separated, the target is defined as the attainment of a material purer than the feed solution, and perhaps also more concentrated. When we say purer, we use this term in a very general sense, i.e. containing a smaller relative concentration of some components.

The target of an extraction process need not be purification; thus the promotion of a reaction, leading to the formation of a desired end product starting from components present in the feed, may equally well fall into the scope of LLX.

In Table 9 desirable targets and practical cases of such target attainment have been linked.

Let us assume that the basic analysis has already been made and that we know that certain separations are attainable by exploiting certain characteristics of the components present, so as to cause selected components to separate, by greater or lesser transfer to a second liquid phase, and that the type of second liquid has been defined according to what it is expected to achieve. How do we now set about delineating a scheme for achieving our aim, and how do we test the scheme so as to arrive at maximum certainty, with minimum expense and in the shortest time?

If we limit ourselves to LLX transfers in order to achieve separations we can consider a number of possibilities, according to direction of transfer:

Pre-extraction	Aqueous → Solvent
Extraction	Aqueous → Solvent
Scrub	Solvent → Aqueous
Stripping/back extraction	Solvent → Aqueous
Solvent clean-up	Solvent → Aqueous Aqueous → Solvent

There is thus only a limited number of procedures whereby separation of components can be attained with an integrated single-solvent system (i.e. where the solvent itself has a constant composition throughout, on solute-free basis). Pre-extraction, which will be aimed at removing components with highest distribution coefficients, is one such procedure. Naturally components with lower distribution coefficients will be co-extracted but to a

TABLE 9. Targets.

Water recovery	Desalination
Water purification	Trace element removal
	Organics removal
Water removal	Concentration
Recovery and concentration	Copper
	Uranium
	Phosphoric acid
Purification	Phosphoric acid
	Citric acid
Preparation	Potassium nitrate
	Caprolactam
Separation	Aromatics/aliphatics
	Aliphatic acids
	$MgBr_2/MgCl_2/CaCl_2$
	$CaBr_2/CaCl_2$
	Ni/Zn

lesser degree. The pre-extract thus obtained must be processed further in order to recover co-extracted components, to the extent that this is required.

The next procedure is a purification or back wash or scrub step. This is useful for cases where, even though the extraction is designed for recovery of the component of interest, nevertheless it is accompanied into the solvent by less desirable components with lower distribution coefficients. Purification or back wash is a selective transfer from solvent back into aqueous phase. The aqueous phase may be a recycle stream from the aqueous strip or a different stream according to conditions. Thus for example the raffinate from the main extraction can be used if the minor component has been extracted by virtue of the transfer of the major components, i.e. mainly at the concentrated end of the countercurrent extraction battery. The scrub stream will always be directed back upstream, e.g. back into the main feed, provided concentrations are not unfavorable.

Thirdly, a back extraction from product can be considered, e.g. if a non-extractable component of the aqueous feed has influenced the extraction of the component of interest which in turn has caused a minor component to be extracted. In the absence of the non-extractable component of the aqueous feed, that is by virtue of it not being present in the product, the back extraction of the major component will be reduced but not necessarily so in the case of minor components.

What has been said above can be summarized essentially as follows: a flowscheme for attaining separation and recovery using LLX transfer as the basic procedure must be built around differences in distribution of the relevant components back and forth, from light to heavy and heavy to light phase in whatever order appears suitable or desirable. Expedients to increase these differences at any stage and in any direction of transfer can be applied provided they are acceptable within the constraints of the particular case.

Whenever extraction is being evaluated it is necessary to determine which end of the transfer system will be controlling according to the distribution behavior alone and in combination with the other components present in the system. This is particularly relevant when the reduction of co-extracted contaminants is being considered; as an example, there are cases where the high concentration of component A influences the transfer of component B; hence separation cannot be postulated on the basis of the separate cases, which will not show up this interaction.

In cases where physical properties like, for example, temperature can influence the distribution coefficient, the conditions for separating co-extractable materials may be quite different. Thus pre-extraction can be done under temperature conditions different from the main extraction, or the scrub at a temperature different from the other steps. In some cases it may be expedient to use a non-LLX approach for removing minor components, to eliminate them from the product, by adsorption, ion exchange, precipitation, etc.

Let us, in the first instance, restrict ourselves to an aqueous feed and to an aqueous or solid product. Most liquid–liquid extractions will entail two mains steps, i.e. transfer from an aqueous phase to solvent and back from solvent to another aqueous phase. Transfers are concentration controlled, and the overall transfer can be expressed as a function of concentration and quantity.

The efficiency of transfer will depend also on the number of equilibrium stages available for transfer. However, in a multi-component system if it is possible to work at an invariant point, the number of equilibrium stages can be eliminated from consideration, since a single stage will be sufficient to ensure the desired result.

A comparison of the possible level of transfer of two components, separately and together, will help to define the format of the separation flowscheme, taking due regard of the aims.

At the early stage of R & D one cannot anticipate being in true steady state with respect to a complex system in all its aspects; nevertheless it is necessary to arrive at a steady state picture in order to facilitate decisions and in order to identify the grey areas needing further clarification. A continuous, operating model of a flowscheme can rarely be considered early in the work, hence other routes have to be sought to attain the information needed.

Single stage-limiting conditions, cross current, countercurrent are all well defined procedures which lead to an overall understanding of the system and hence are directly relevant for devising the flowscheme. It is clear, of course, that questions such as minor component build-up have to be studied using a "distorted model" if a picture is to be attained which can be relied upon. Limiting factors in transfer may be listed as: product yield; product concentration; product quality; solvent losses.

The Two-Liquid Phase System

Not only is the two-liquid phase system useful for actively separating components, but it can also be applied for keeping specified components separated, while permitting transfer of other components across the interface. This approach has been well exploited in phase transfer catalysis. Recently other examples have come up. One such example is presented in Fig. 5.

Another use for the two-liquid phase system is to supply reactants to a system by dissolving them in separate liquid phases, selected so that one component can distribute, but the other cannot; the product of the reaction

$$NaCl + H_2O + CO_2 \rightleftharpoons NaHCO_3 + HCl$$

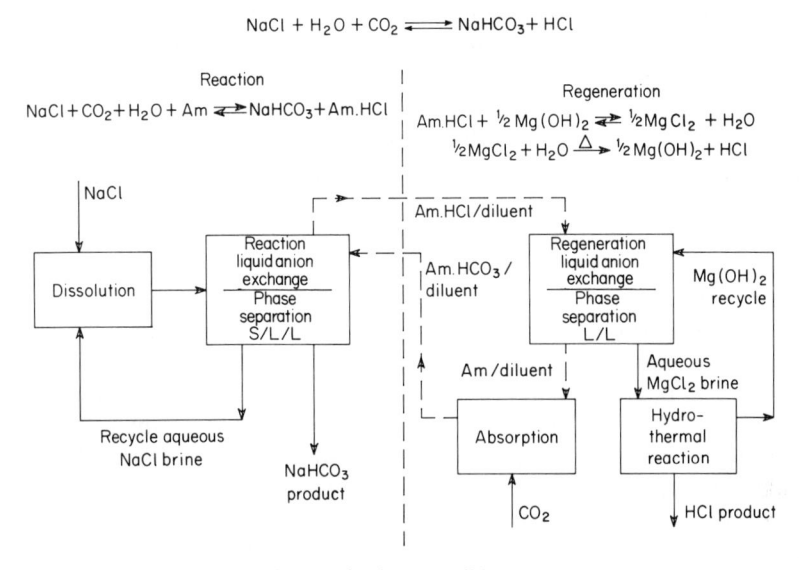

Reaction:
$$NaCl + CO_2 + H_2O + Am \rightleftharpoons NaHCO_3 + Am.HCl$$

Regeneration:
$$Am.HCl + \tfrac{1}{2}Mg(OH)_2 \rightleftharpoons \tfrac{1}{2}MgCl_2 + H_2O$$
$$\tfrac{1}{2}MgCl_2 + H_2O \xrightarrow{\Delta} \tfrac{1}{2}Mg(OH)_2 + HCl$$

Fig. 5 LLX for keeping incompatible components apart

may behave in various ways, e.g. precipitate and collect at the interface, dissolve in the upper phase, dissolve in the lower phase, precipitate from the lower phase. If there are two components in one of the phases, both of which react with the reactant supplied by means of the second phase, but they and/or their reaction products have different characteristics as regards wetting or distribution, this can also lead to separation of components. This then becomes a multi-phase system including also depositing solids.

This approach, in principle, has been used in several cases, strangely enough in as widely different spheres as inorganic chemical processing and biotechnology separations, but the scope has not been defined nor the possibilities analysed.

The Phase Rule in LLX

In an equilibrium-limited multi-phase, multi-component system, the Gibbs Phase Rule is an invaluable tool since it reduces complex systems to dimensions that can be handled. The significance of the Gibbs Phase Rule for LLX was not spelled out initially, being of little interest in the study of solute–reagent interactions. However, as applied researchers needed limiting cases and simplifying assumptions in order to be able to dominate their complex systems, they were able to appreciate the simplicity of a system controlled by the Phase Rule.

The beauty of the application of the Phase Rule in LLX systems can be exemplified for an acid/salt system where the anion is multivalent and therefore permits the formation of a series of acid salts. Each salt will have a fixed composition, but the conjugate aqueous phase may not be fixed since there will be some span of acidity in the liquid phase for each of the acidic salts as solid phase. Let us use the ternary isotherm in Fig. 6 as a case in point. If the acid crystallizes, then ab spans the pure acid area, bc spans the pure mono-basic salt, cd spans the pure di-basic salt, de spans the pure tri-basic salt, b represents mixtures of acid and mono-basic salt, c represents mixtures of mono- and di-basic salts, d represents mixtures of di- and tri-basic salts.

Consider now a system containing also a solvent which can extract acid, say a C_5 alcohol. When an aqueous phase of b composition is brought in contact with the solvent in the presence of the mono-basic salt, the solvent will extract acid along the line bc, but the mono-basic salt remains the single solid phase. At point c, the solvent will extract acid converting MH_2Z to M_2HZ, but the aqueous phase will remain constant at point c. The system is therefore self-controlling, remaining at some point, according to what is being fed into the system. This type of weakly basic, solvating solvent, however, will only operate in the acidic range, i.e. provided M_2HZ is sufficiently acidic. If it is basic, then a solvent of sufficient basicity must be used in order for the decomposition or acid extraction (or abstraction) to take place.

Another interesting example is the point at which a second solvent phase separates. Thus at fixed temperature, in Fig. 7 abc is the zone of three liquid phases, hence any point inside the triangle abc will separate into these three fixed composition phases, so here again the system is self-controlling.

Yet another example, as shown in Fig. 3.3, is the metathetic reaction $MA + \overline{HB} \rightleftharpoons \overline{MB} + HA$; at fixed temperature and acidity, with both MA and MB present as solids, this is invariant, hence MA can be converted quantitatively to MB by extraction of the stoichiometric amount of HA, provided MA and HB are fed in the stoichiometric ratio; the aqueous phase will not change in composition during the extraction and this fixed composition is independent of the quantity of aqueous phase present.

If a gas or vapor phase is present, this can be taken into account too, following the Phase Rule and calculating the degrees of freedom. When desired, certain parameters can be fixed, thus removing a degree of freedom and ultimately reducing the system to a pseudo-invariant one.

When solvent systems are being compared for initial selection, it is often difficult to select conditions which are comparable. In this case "reduced" parameters can be used where applicable, as is accepted in comparing materials generally, or more usefully pseudo-invariant points can be devised so that the system becomes fully comparable, within the limits of the Phase rule. Thus, for example, comparable partial water vapor pressure of the

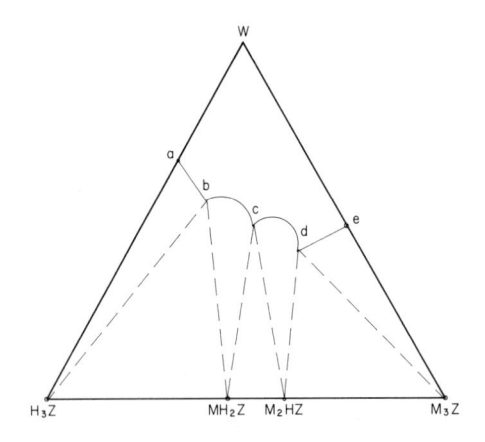

Fig. 6 Ternary diagram—water/acid/salt system

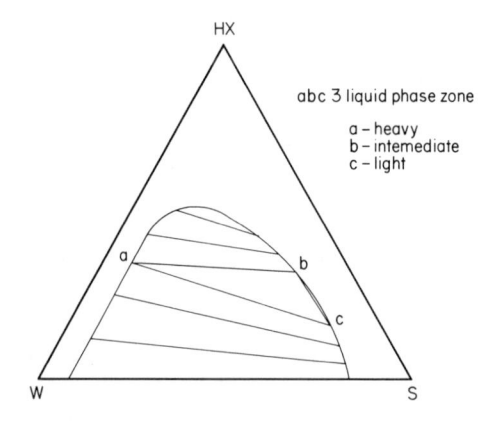

Fig. 7 Ternary diagram—acid/water/solvent. Isotherm with 3-liquid phase zone

aqueous phase can be used, corrected possibly for solubility of solvent in the aqueous phase, or saturation with respect to a specific solid component can be selected as a fixed reference point. Two different solvents at equilibrium with the specified aqueous phase, whether at fixed partial water vapor pressure, or in the presence of the solid phase, are themselves, as it were, in equilibrium although not in contact, since they are separately in equilibrium with a common aqueous phase, say at a pseudo-invariant point. This procedure, though clearly not rigorous, is extremely helpful and saves considerable testing time and effort. Thus when a full flowscheme has been worked out for a particular solvent, the critical points can be selected by inspection and used for comparison with a different solvent candidate; the possible influence on the flowscheme can be forecast as a first approximation by evaluating the two solvents at the selected points of comparison.

Let us take a very simple case. Supposing the water balance in a process constitutes a critical aspect, then the solvent selection will of necessity have to consider the extent of water transfer taking place during the extraction/stripping sequence. Instead of having to develop the whole phase diagram for each solvent, it will be sufficient to base the first comparative evaluation of solvents on behavior at selected points in the scheme, provided these points are invariant so that they are true points of comparison.

Another interesting aspect of working at invariant point is that since the aqueous composition is fixed, so will the conjugate phase be defined for each solvent; hence a single contact will suffice, making countercurrent or multiple-stage contacting unnecessary. Returning to the salt system in Fig. 6, if MH_2Z and M_2HZ are present together, point c will be equivalent on contact with any solvent irrespective of the solvent composition even though c itself may move on the diagram of a specific solvent, i.e. when the additional dimension is included for the additional component and the additional phase. The classic approach to using phase diagrams, developed for multicomponent salt systems, relied on using projections, that is the water axis was eliminated, basing only on "dry basis" comparisons. Similarly here we are proposing to eliminate the solvent axis, hence the aqueous solubility diagram becomes a valid reference and comparison point.

Diluent/Extractant Interactions

Diluent/extractant interactions have a marked effect on distributions in solvent extraction and liquid ion exchange systems. The influence of diluents on the extraction of acids by high molecular weight amines has been extensively examined. Also the solvating ability of amphiprotic solvents such as alcohols is well known, and they have been used as additives in some cases; thus alcohols alone are as effective as nitrobenzene alone in enhancing the base strength of tertiary alkyl amines; furthermore, mixtures of nitrobenzene and alcohol show a non-additive (synergistic) effect on the apparent formation constant of tertiary amine hydrochlorides. The effect of protogenic diluents can be exploited also to change the "normal" order of selectivity of tertiary amines for anion exchange, and the equilibrium in metathetic reactions with quaternary ammonium salt ion exchangers can be shifted by addition of strongly protogenic diluents. Alcohols added to cation exchangers in "inert" hydrocarbon diluents cause pronounced changes in the "acidity" of the ion exchanger. This can be seen in Fig. 8. Ketones and ethers have similar effects.

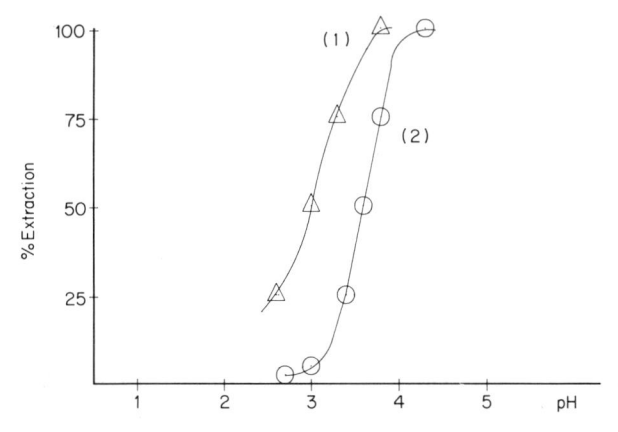

Fig. 8 Effect of diluent/modifier on cation extraction (1) kerosene diluent (2) 50/50 vol% n-hexanol modifier/kerosene diluent

The effect of reagent/diluent interaction in the extraction of metal complexes in amine sulfate systems shows extremes in behavior, for example a primary amine in a non-polar diluent will extract Fe^{3+}, while a tertiary amine, even in a polar diluent, extracts only acid. Combinations can be selected to suit specific requirements.

Multi-component Multi-phase Systems in LLX

In its broadest sense LLX can be classified together with volatilization and precipitation as a separation technique entailing mass transfer across a phase boundary. Since it is essentially a reversible transfer system, at equilibrium, any single phase can be selected to describe the system as a whole.

By definition, solvent extraction requires that two liquid phases be present but there is no other limit *a priori* on the number of phases; a limitation is placed, however, by the Phase Rule defining multi-component multi-phase systems.

The rigorous definition of a multi-phase/multi-component system in LLX is always difficult and often quite impracticable. Fortunately, however, such systems can usually be manipulated towards specific aims, without the need for such rigorous characterization. On the other hand, if the system is invariant within the meaning of the Phase Rule, any single phase at equilibrium defines the system completely. In the same vein therefore it is possible to make a system pseudo-invariant by preselecting the level of specified parameters and then by inspection to select one phase as controlling and so to characterize the whole system. The examples of this approach cover a wide range from metathetic reactions, through ion exchange, to concentration of brines by water transfer, and so on. It seems therefore that a more generalized approach than the customary one to solvents and to multi-phase/multi-component LLX systems is possible and may permit the field of extraction to be expanded.

While the range of solvent candidates is large, the types of organic compounds actually used as bulk solvents are relatively limited. In inorganic chemical processing, LLX applications have been classified either according to the object (e.g. copper extraction, phosphoric acid purification), or according to the operation (e.g. ion exchange, metathetic salt/acid reactions, separation of metal values, etc.).

All these cases have in common the fact that an aqueous solution constitutes one of the phases, and that the second liquid phase consists of, or contains, an organic component which is loosely defined as the solvent or extractant. As a result of this combination, in some cases the degree to which water itself distributes between the phases becomes of considerable significance in defining the characteristics of transfer, and in determining the separation that can be attained. Thus the activity of water may be not less important a parameter than the concentrations of the solutes nominally being separated in the specific extraction process. From this it follows that water can be regarded as essential in the mechanism of transfer between the phases. Indeed, spectacular separations can be achieved on the basis of differences in transfer of hydrated species between an aqueous phase and an organic solvent which has a capacity for dissolving water. This can be exploited to great advantage for separating salts from brines and separating salts from one another. Examples show that separation factors of an order of magnitude between salts can be attained and that preference of extraction can even be reversed, depending on the solvent system chosen.

Since water is such an effective solvent for ions and polar molecules, the degree to which water itself distributes between the phases may be of considerable significance in defining the characteristics of transfer, and in determining the separation that can be attained.

One measure of the extent to which water can be transferred to or from a system is the partial pressure of water in the system; this also defines the thermodynamic activity of water in that system. A very elegant and original exposé of this has been presented within the framework of utilizing LLX as a means of concentrating brines without evaporation. The significance of solvent polarity in solvent extraction is well accepted, yet its exploitation in practice, for inorganic processing and separations is very limited. All in all, the solvent types which have found practical application are very few. Groups of polar compounds seem to have been overlooked. This is exemplified by the special characteristics and solvent potential of classes of compounds like amides, glycol ethers, organic carbonates, esters and so on.

Furthermore, there are so many reactions and separations to be achieved by exploiting singular characteristics of particular multi-component/multi-phase systems provided they contain at least two liquid phases which thus qualify them as candidates, by definition, for LLX separations.

Once we accept the premise that LLX is a separation technique based on transfer across a phase boundary, it follows logically that all expedients for favoring such transfer are legitimate, provided that technological and economic feasibility are not sacrificed in the final instance. This also leads logically to the inspection of systems to identify controlling parameters for exploitation in mass transfer.

Interaction of water vapor pressure and water content in conjugate solvent phase is shown in Table 10 for several salt/water/solvent systems. In Table 11, too, the relationship of water in solvent phase to water vapor pressure of the conjugate aqueous phase can be seen.

When there is no special bonding between the solute and solvent, i.e. only water transfers across the boundary, then the water levels in various solvents reflect the ionic strength of the aqueous phase and can be used as a measure of the partial water vapor pressure in brines. When there is a specific interaction between solute and solvent, the water activity alone no longer suffices for defining the equilibrium system, H_2O/solute/solvent, as a whole. At the same time the distribution coefficient of the solute which relates to its concentration in the conjugate phases also does not define the system. In a labile interacting system of a multiple number of components and phases, the distribution coefficient may lose its practical usefulness unless some auxiliary device is applied for defining the state of the system.

TABLE 10. Water content of butanone, butan-1-ol, butan-2-ol, in equilibrium with saturated solution of salts at 30°C.

Solute	Vapor pressure (mmHg)	Butanone	Butan-1-ol	Butan-2-ol
		(g H_2O/100 g solvent)		
None	31.8	14.2	26.0	53.6
K_2SO_4	30.8	13.3	20.7	34.8
KNO_3	29.2	8.5	15.5	21.7
KCl	26.6	6.3	10.4	11.2
NH_4Cl	24.5	4.9	9.3	10.6
NaCl	23.9	4.1	7.4	8.7
NaBr	19.7	2.4	5.7	5.3

TABLE 11. Vapor pressure of water as a measure of water in solvent.

Systems: K_2CO_3/H_2O/cyclohexanol
$MgCl_2/H_2O$/cyclohexanol
Temperature: 40°C

	Aqueous solution (wt % solute)	Vapor pressure H_2O (mmHg)	Solvent phase (wt % water)	(cations meq/g)
K_2CO_3	zero	55.3	11.7	—
	30.8	47.6	6.1	—
	42.4	38.3	4.1	—
	51.8	27.6	2.3	0.005
$MgCl_2$*	23.5	42.0	5.1	0.03
	34.2	25.0	6.3	0.54

*The measurements were made with natural brines containing also $CaCl_2$ as a secondary component.

TABLE 12. Two- and three-liquid phase systems.

a. Acid/water/ether
Temperature 30°C
all compositions as weight per cent

System	Overall composition	Aqueous Bottom	Organic Middle	Top
H_2SO_4	50	63		3
H_2O	27	33		—
n-Butyl ether	23	4		97
H_2SO_4	51	63	36	4
H_2O	24	32.5	10	—
n-Butyl ether	25	4.5	54	96

b. $NaCl/HCl/H_2O$/tertiary amine

Trilauryl amine	Saturated	25	
Trilauryl amine.HCl	with NaCl	25	
Pentan-1-ol		50	
Trilauryl amine	Saturated	12	85.5
Trilauryl amine.HCl	with NaCl	37	1.5
Pentan-1-ol		51	13

Here the Phase Rule, with its definition of degrees of freedom and invariant compositions, can be utilized to great advantage. Even in the simplest system, as described in relation to water transfer, the invariant aqueous solution in the presence of its solid phase defines a real limit of transfer, and this provides a basis for comparison, either among solvents or among aqueous phases.

Systems in which one aqueous and two solvent phases are present become invariant if the degrees of freedom are reduced by fixing selected parameters such as significant concentrations and temperature.

The separation of a second solvent phase constitutes a second liquid/liquid interface for mass transfer, hence by definition, it provides a more sophisticated separation system, and opens up a whole field for study.

Two cases which pass from a two-liquid to a three-liquid zone are given in Table 12.

The "Second Liquid Phase" — Freedom of Choice

The unit operations that entail separations are of absolutely general applicability, provided the physico-chemical requirements of the operation are satisfied. Thus crystallization can be considered for any material which will form a solid phase; similarly precipitation is close to crystallization provided solubility permits; filtration can be considered whenever solids must be separated from liquids; evaporation provided vapor–liquid equilibria favor the purpose, although aids can be used to promote the step being considered, e.g. like adding an azeotrope former to help achieve the vapor–liquid equilibria which favor the evaporation required.

Different from these are adsorption, absorption and LLX, since these inherently require the introduction of an "aid to separation", apart from the energy balance which must be satisfied in all cases.

This "aid to separation" will be an integral part of the operation in principle; thus in LLX we have the "second liquid phase" without which, by definition, there is no possibility of LLX; in adsorption we have the "adsorbent" and in absorption the "absorbing phase". In all these cases, the "aid to separation" is assumed by definition to suit the case in mind, without any need to define the mechanism of transfer. It is clear, however, that the mechanism of transfer is the determining factor even in a blind "trial and error" system. In LLX, the requirement is distribution, hence the distribution coefficient is the measure of successful choice of the second liquid phase, but it says nothing about why there was success or how to seek success.

An analysis has been made of the mechanisms of transfer in LLX, and these have been defined as: solvation, ion-pair formation, complexation and, of course, straight partitioning as a function of solubility in the two phases, following Raoult's law. The first three imply the use of a second aid—the "solvating agent" "the counter ion", "the complexant" which we shall call here the "transfer agent".

The necessity to have a transfer agent may be considered a complicating factor on the one hand, but on the other it extends the freedom of choice and facilitates separations since it permits the true mechanisms of transfer to be selected by analysis of the nature of the material(s) to be transferred.

LLX is based on the partitioning or distributing of a material between two liquid phases in contact. In the creation of the two-liquid phase system, there are two participants—the "given" and the "chosen". Were it not for the "chosen", LLX would be of extremely limited applicability, since systems which naturally consist of two liquid phases are certainly rare, and materials which distribute as required simply on the basis of relative molecular solubility are also limited. The choice of the second phase can be regarded as limitless in scope, from the basic point of view of distribution. However, distribution is not enough to satisfy separation and recovery, both of which are needed if the LLX is to have viability. This viability will derive from the possibility of "extracting" into a second liquid phase, and "stripping" from this phase. Essentially therefore, reversible transfer is required. In the execution of free choice for the second liquid phase, this constraint of reversibility is a primary factor when a process is under consideration. At this point, we can say categorically that LLX requires transfer "through" a second liquid phase—this in no way confines the second phase as to quantity, nor the system as to mode of contact. So the use of second liquid impregnated resins or of second liquid membranes is no less LLX than the use of a bulk of second liquid.

Something has been written about where to search for aids to choice of second phase with or without a separate transfer agent, and the strategy to follow in seeking such aids. Thus, there is no doubt that the basic chemical background on complexants and the use of these made in analytical chemistry were the source of choice of transfer agents—hydroxy oximes, etc.—for copper separation and recovery, and that without this it is unlikely that that great success story would have come to pass.

In considering how to make a choice of second phase, two degrees of separation are to be considered—bulk separation and fine separation. Take the case of copper once more: bulk separation relates to "recovery" of the copper, at the low concentration at which it is present in the dilute aqueous leach liquid, while fine separation relates to "quality", that is, to the degree of separation from contaminating materials—the more similar these contaminants are to the main material, the more difficult the separation, and the greater the importance of transfer agent choice and/or choice of conditions in transfer.

The choice of the second phase must therefore take into account the ease of transfer "in" and the ease of transfer "out". In certain cases, it may be valid to combine two types of operation, e.g. LLX and evaporation, LLX and neutralization, LLX and precipitation. When it is to be LLX and LLX, it is necessary to choose the second phase so that it is amenable to the application of some parameter change in order to achieve the desired transfer.

Parameter changes which have been exploited fairly widely are imposed pH control and temperature. A change which has been suggested but apparently not exploited is the modifier:extractant:diluent ratio, in a system during extraction and stripping. The second phase will be chosen according to the sensitivity it displays in relation to the parameters selected for change.

The practical distribution coefficient is the resultant of all the equilibrium constants which apply in a particular case, hence the recognition of the relevant individual constants and the likely effect of a parameter change on these constants will help to identify the constants that are controlling, and hence will be the ones to be adjusted so as to accomplish extraction as well as stripping satisfactorily.

In Situ Self-Generation of Two Liquid Phases

Liquid–liquid extraction for material separation is an active, transitive operation; in other words two liquid phases are brought into contact so that material can transfer across the interface, hence "extraction" from one phase into the other is promoted.

In situ self-generation of two liquid phases for material separation is an intransitive operation in the sense that the distribution of material across the interface occurs by virtue of the system's passage from an homogeneous to an heterogeneous equilibrium state. All components present in the system participate in the passage simultaneously* hence the transfer of interest does not lag behind the passage from homogeneous to heterogeneous state. In a multi-component system, at the critical or plait point, two phases in equilibrium approach the same composition; as one moves into the heterogeneous zone the compositions of phases in equilibrium will be increasingly different.

*The nearest analogy, perhaps, would be vapor–liquid equilibrium in an infinitesimally small closed system, by abstracting heat from a vapor phase.

The *in situ* generation of two liquid phases for separation and recovery of a component or components from a multi-component system differs in a significant aspect from the normal LLX pattern, which consists of contact, transfer in one direction, phase separation, followed again by contact, transfer in the reverse direction, phase separation, and repetition of the cycle. In the *in situ* case, as it has been practised thus far on a small scale, the conditions are set up for the transition from homogeneous to heterogeneous zone, with all the simultaneous transfers that this entails, followed by phase separation but not necessarily a cycle operation.

A practical case of *in situ* generation of two liquid phases was presented in LLX for water desalination, using organic amines which had a large capacity for water dissolution, but were themselves only very slightly soluble in an aqueous medium. It was the amine–water phase which was then used for generating heterogeneity by a minor temperature rise; the two equilibrium liquid phases thus generated had extremely different compositions, i.e. they were as far as practicable from the plait point. Essentially this latter step could be regarded as occurring in a two component system, since salt distributed to a very minor degree; the relationship was, therefore, similar to a temperature/mutual miscibility relationship for the two components, water/amine.

In looking now at bipolymer/aqueous systems, or polymer/salt/aqueous systems, conceptually it would be convenient to regard these also as consisting of two essentially immiscible polymer components with water distributing between them. A fourth component (i.e. not the initial two polymers and water) would distribute between these two phases according to the interactions in the system. This mode of approaching bipolymer systems may aid in arriving at cyclic operations as was indeed the case in the water desalination study.

LLX As Tool for Achieving Transformations

Whereas the application of LLX for recovery or purification is widely accepted and implemented industrially, its application for transformations has been limited. Nevertheless there are a number of processes in which LLX is utilized as the central unit operation for promoting the desired transformation. Immediately several general case types come to mind, one where the LLX step promotes the transformation by removing a co-product or main product from the reaction zone, another where the presence of the second phase permits keeping incompatible reactants separated from each other. A third case is where the reaction takes place in a "stripping" step.

Thus, for example, when an acid has been extracted, the strip may be carried out by a base to give a salt in aqueous solution as product. Examples are summarized in Table 13.

Possibly the reason for the limited application of this approach to LLX is that transformations have not been classified in this manner, and/or that modern chemical industry places its stress on kinetically controlled reactions more than on simple metathetic, equilibrium controlled reactions. However, in kinetically controlled reactions too, e.g. where consecutive reactions are possible, hence the concentration of product A should be kept low so that the rate of A → B will be low, it would seem that LLX, for removing product A as formed, should be ideal.

LLX is best suited to continuous operation, which fits in well with the system wherein a component is to be kept at some specific level by continuously topping off an increment as formed.

The beauty of this approach is that full separation of products from either phase is not mandatory, since return for recharging or replenishing would be the accepted regime.

Using a second phase as a mode of keeping incompatible components from interacting has been described to a limited extent within the framework of LLX. On the other hand, phase transfer catalysis which is not generally

TABLE 13. LLX as a tool for transformations.

Promotion of equilibrium reactions
  e.g. $KCl + HNO_3 \rightleftharpoons KNO_3 + HCl$

Recovery of concentrated product from dilute solution
  e.g. by two ion exchange steps.
  Dilute Cu^{2+} extraction → $\overline{R_2Cu}$
  Acid + $\overline{R_2Cu}$ recycle stripping → $Cu^{2+}_{(conc.)}$ + $\overline{HR}$

Salt production by base stripping of ion pair extract
  e.g. $\overline{R_3NHX} + M^+OH \rightarrow \overline{R_3N} + MX + H_2O$

Dissociation extraction for separation of acids of different pK's keeping incompatible components apart
  e.g. $NaCl + CO_2 + H_2O \xrightarrow{Amine} NaHCO_3 + \overline{AmHCl}$
  $\overline{AmHCl} + \tfrac{1}{2} Mg(OH)_2 \rightarrow \overline{Am} + \tfrac{1}{2} MgCl_2 + H_2O$

Disproportionation of double salts using solvating solvents
  e.g. $K_2SO_4 \cdot H_2SO_4 \xrightarrow{C_5OH} K_2SO_4 + \overline{H_2SO_4}$

connected with LLX, but which is now fully accepted in organic and petrochemistry, does exactly this thing, although possibly in the reverse. This is a highly active area which can furnish data and background to lead to LLX processes.

The large number of two-phase liquid systems that are known, and the many expedients whereby they can be modified almost in a continuous manner, also the wide range of solubility and selectivity that can be covered in a continuum of graded pairs of liquid phases, makes one assume that no matter what constraints are imposed, a suitable system can be found for any desired case. Separations that can be achieved by other means may very likely also be attainable in two-liquid phase systems. (This does not imply that the separation so performed by LLX will necessarily be viable or practicable, though possible.) Since liquid–liquid systems have critical solution points, it is often possible to generate two phases at will—to be exploited for promoting reactions with concomitant separations.

Extension of LLX — Where and How?

There has never been the need to consider how to extend distillation or where to introduce it, since the use of distillation is self-evident in technology whenever it can be applied. The accent has been therefore on technological or economic improvement. In this regard LLX is quite different. The application of LLX is never self-evident, nor is it inherent in any system. Its application must be intentional, after evaluation, comparison and selection. In a sense, therefore, the extension of LLX requires considerable ingenuity since the system has to be created, and it also requires appreciation and awareness of scope and possibilities, and a knowledge of the interactions of systems in a broad manner. The case of separation between the two acids HCl and HNO_3 is a model which is rare in its characteristics though not widely known. Essentially it is modelled on distillation, being absolutely analogous to a system entailing stripping and rectification, but with the separation being from the non-aqueous phase. Another such case utilized on large industrial scale does not readily come to mind. It is usually customary to seek selectivity or specificity so as to attain the main separation in the primary extraction, but in the HCl/HNO_3 separation, where there is the by-product (HCl) and the feed (HNO_3), no attempt is made to separate between HCl and HNO_3 in the primary step; on the contrary, it is clear that with solvating solvents such as alcohols, NO_3^- always extracts preferentially over Cl^-, which also immediately makes it obvious that separation of Cl^- from NO_3^- must be based on a rectification analogy. For the analogy,

I. Liquid–Liquid Extraction (LLX) Procedure—An Integrated Approach 47

therefore, one must regard the solvent containing $\overline{NO_3^-}$ and $\overline{Cl^-}$ as analogous to the vapor phase in rectification, thus necessitating the return of an aqueous nitrate phase as the countercurrent phase for rectification similar to product condensate in distillation. The aqueous phase out of rectification will subsequently be stripped of NO_3^- by feeding acid-free solvent in countercurrent which will generate a phase similar in composition to the feed phase which it will join. This is presented in Fig. 9.

In general, therefore, if we have an aqueous feed containing a material A with relatively high distribution which is *not* the desired product, and a second material B of lower distribution which *is* the desired product, ideally the extraction could take place with S containing a quantity of A in recycle, so as to extract B and then to separate the two on a basis similar to the above, as shown in Fig. 10.

The recycle of A to rectification compensates for A extracted so that (A) out of rectification into extraction is kept constant, and A into extraction and out is kept constant. What the system requires therefore is a fixed quantity of A in constant recycle to accept the B being separated initially. Volume of S will be determined by the K_D's of the two materials and by the ratios of the K_D's according to the direction of transfer.

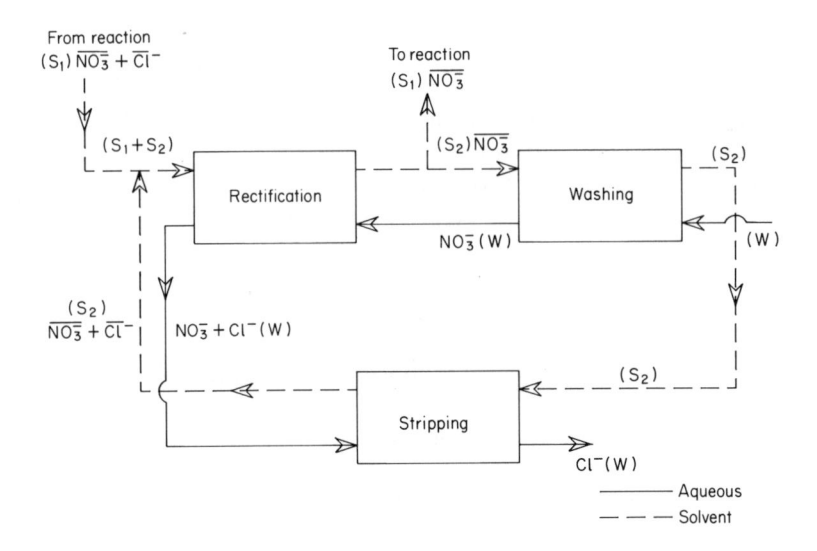

Fig. 9 HCl/HNO$_3$ separation

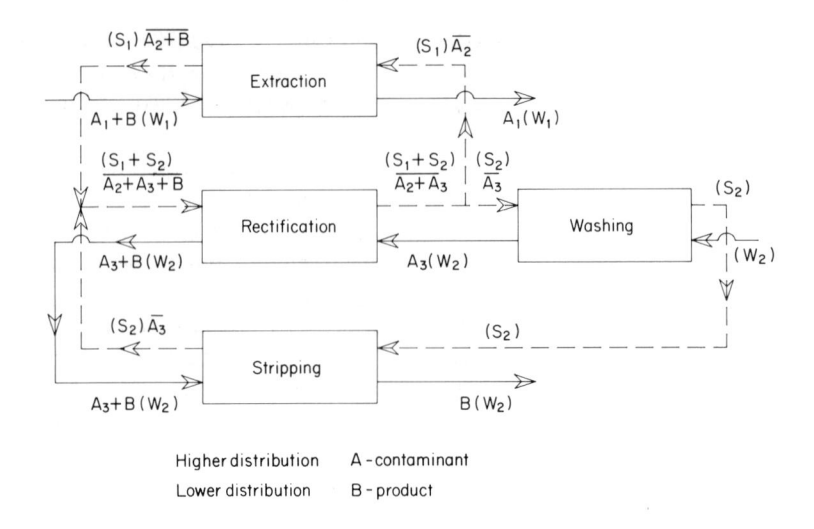

Fig. 10 LLX rectification/stripping analogy

The case described for HNO_3/HCl is based on a solvating solvent, the same approach adapted to each case can be used for ion pair or ion exchange extraction. In the latter case the relative pK's will be determining. The level of recycle of the preferentially extracted component will depend on the equilibria attainable at the inlet end of the rectification, and the number of stages, exactly as the reflux ratio in distillation is a function of V/L equilibria and the number of rectification stages available.

Application of a similar approach is followed when the desired material has the higher distribution; here scrub or back wash with an aqueous phase is used for separating the extracted less favorably distributing component. This use of the technique is fairly widespread in LLX applications; it is in fact similar to the rectification concept above, since the scrub is done with an aqueous phase which may or may not contain a quantity of the pure product so as to preserve the level of concentration in the extract in and out of scrub.

In a system in which two very similar materials are being extracted so that the separation in LLX becomes too cumbersome because of the small preferential distribution, it may be necessary to combine technologies for the final separation, utilizing a property different from liquid-liquid distribution for separation, e.g. relative volatility or relative solubility. It may also be possible to delay the LLX separation to some point farther along, e.g. after concentration or crystallization of main product.

It is clear now that for a new separation, i.e. between materials perhaps not previously separated by LLX, it is legitimate to visualize various procedures before selecting a model for the specific case under consideration.

Biotechnology

Until fairly recently, LLX was not a separation procedure of choice for bioprocesses. However, "biphase aqueous systems" have been applied on a small scale since the 1960's, but not in the same sense as LLX has in hydrometallurgy, petrochemicals, pharmaceuticals, organic and inorganic processes, that is to say not in a steady state cyclic context. On the other hand, the efficient separation, recovery and purification of materials produced by microbiological procedures becomes increasingly important as scale of production increases.

Two classes of materials should be considered when reviewing the possibilities of applying LLX in bioprocesses. The first type covers low molecular weight specific materials such as acids, modified sugars, hydrolysed protein products. The second type covers high molecular weight materials such as enzymes and proteins. The former class is much nearer to the kinds of materials encountered in chemical technology, hence separation and recovery can be based on analogy with established separation techniques and analytical procedures. The latter class presents a more difficult problem because of the sensitivity of these materials to their environment and the danger of inactivation and for denaturation.

In bioprocesses, it can be assumed that the primary environment will be aqueous, i.e. that the materials will be generated in an aqueous environment (not necessarily homogeneously dissolved, however) from which they are to be separated, recovered and purified.

In chemical technology, a striking aspect of LLX is the possibility of devising systems with a high degree of selectivity for desired materials, and a high degree of rejection of others.

Within the framework of unit operations, there are two approaches to separation and recovery; the one relies on removal of water, the other on abstracting the product. In biotechnology, both approaches have been followed and the unit operations used will be removal of water by distillation for smaller molecules or by membrane permeation techniques such as ultrafiltration or reverse osmosis, or the separation of the product by precipitation, ion exchange, or crystallization.

Liquid–liquid extraction has been utilized only to a limited extent. However, regarding acidic or basic materials of low molecular weight, there is full analogy with comparable materials of non-bio-origin. There is, therefore, no reason not to consider LLX in such cases. Acids can be recovered by weakly basic solvating solvents or by strongly basic reagents, while basic materials can be recovered by cation exchange or by ion pair formation. The use of complexing reagents can be considered equally well, but suitable ligands need to be sought.

In the case of high molecular weight materials, the analogy is not so close, mainly because of the constraints regarding the ambience.

Biphase aqueous systems, generated by mixing two aqueous polymer solutions of different molecular weights or solutions of polyelectrolytes, have been utilized for enzyme separation and recovery, in the presence of suitable buffer systems.

Precipitation, by using complexing or ion-pair formation, has been well established for enzymes, and indeed the use of polyelectrolytes follow this path too, so here also there is considerable analogy on which to base an approach to LLX system design.

This then is the challenge—how to devise LLX systems for transfer and separation of materials in bioprocessing without losing the essential characteristic of LLX technology. This characteristic is the possibility of steady state operation keeping the "transfer phase" in closed cycle. The "transfer phase" would be analogous to the "solvent phase" in LLX in chemical processing. It is the "second" liquid phase into which the transfer occurs from the primary feed across the liquid boundary, and out of which the transfer occurs into the product phase, again across the liquid boundary, thus freeing the "second" liquid phase for return and reuse.

In bioprocessing, adsorption has been exploited considerably, especially in chromatographic separation and purification, both for production and for analysis, hence here, too, there is analogy from which to draw approaches to reagent selection.

However, the challenge in applying LLX technology to bioprocesses does not lie only in devising the two-phase systems for partition or transfer, but also in how to eliminate water and thus to maintain steady state. Thus far, when using biphase aqueous systems, two recovery operations have usually had to be combined, e.g. partition and adsorption, partition and ultrafiltration, partition and precipitation. The problem, therefore, may not relate so much to the materials being separated and recovered, as to the system itself, which must remain in steady state for cyclic operation.

In the case of low molecular weight materials not sensitive to the environment, the second liquid phase can indeed be essentially non-aqueous, so that water entering with the feed stream will be rejected with the raffinate after extraction. Here a two-phase system will be inherent in the characteristics of the second phase and its components, as regards the degree of mutual immiscibility with an aqueous conjugate phase. This two-phase system will not have to be generated by adding something to the aqueous feed, as is the case when polymers or polyelectrolytes of differing molecular weight are used for generating phase "demixing". The need for secondary recovery from the aqueous phase, as a means of avoiding rejection of the reagents, thus becomes redundant.

For high molecular weight materials such as proteins and enzymes which *are* sensitive to the ambience, the design of the two-phase system and the selection of a transfer agent, with suitable product/reagent interaction, becomes the crux of the problem.

It would seem that analogies must be sought in biotechnology itself, where reagents of certain types are in fact used for separation and recovery of enzymes, by complexation or ion-pair formation. Here the technology practised in enzyme immobilization could serve as a source for analogies, by examining the steps involved in immobilization and interpreting what happens. The question will then be whether suitable second liquid phases can be devised to incorporate similar compounds which will draw the enzyme out of the aqueous feed into this liquid phase, comparable to the interaction that permits immobilization into or onto a solid phase. It should be stressed that at this point we are not looking at the second liquid phase as an immobilization agent, but as a separation agent. Hence, it becomes essential for the transfer to be reversible, so that acceptable changes in system conditions can promote transfer back to a fresh aqueous phase, and thus release the second phase for recycle.

Two aspects, therefore, seem to be required for successful application of LLX in biotechnology, and these are rejection of water, so as to maintain steady state, and reversibility, so as to enable a cyclic extraction to be implemented.

When ion-pair formation or ion exchange is the mode of transfer, both requirements can probably be satisfied because of the similarity to numerous processes which employ the same devices. When separation is based on distribution or partition, some additional step may have to be applied in order to achieve water rejection, and possibly also for release of the desired product from the solvent phase.

In certain systems, temperature may be a powerful tool for changing mutual miscibilities, thus leading, for example, to water rejection; this could also be a mode of releasing the solute if it prefers an aqueous ambience. Changing pH is another powerful tool for changing distributions. Unfortunately temperature and pH are both parameters that can usually only be varied within limits in enzyme and protein systems. In the latter case, a different approach must be sought for water rejection and also for product release.

In biphase aqueous systems as are applicable thus far for biotechnological separations of enzymes and proteins there are conceptually two aspects to be considered; the one relates to the generation of the two-liquid phase system, the other to the distribution, separation and recovery of the product.

In the simplest cases, the two-liquid phase system can be regarded as a ternary system (whether this be two polymers and water, or a polymer, a salt and water) wherein there is some mutual miscibility or distribution of all the components. There is no difficulty in determining the phase diagram for such systems under controlled parameters, e.g. of pH, temperature, osmotic pressure etc. and defining the limits of the two-liquid phase region. The aspect that distinguishes these systems from other LLX systems is that in most other cases only the components of the transfer phase need be selected, while the given feed phase is used as such. Even in cases where there is, for example, considerable transfer from feed phase into the selected transfer phase, there is usually no problem in avoiding moving out of the two-liquid phase zone. In biotechnology thus far, it has been necessary to modify the feed phase too in order to enter the two-phase regions. In a once-through system, these aspects present no problem conceptually, the only requirement being that the distribution of interest be strongly in favor of the second phase. It is when one wishes to consider a cyclic steady state operation, as is common in LLX, that one comes up with a problem, since such a cyclic operation requires some form of product stripping from extract. In order to retain the LLX concept, this means generating two phases once more, or using a different operation for releasing the distributed product, so as to liberate the second phase for recycle. If, for example, there is an effect of temperature as shown in Fig. 11, this becomes possible and the flowscheme has the form presented in Fig. 12.

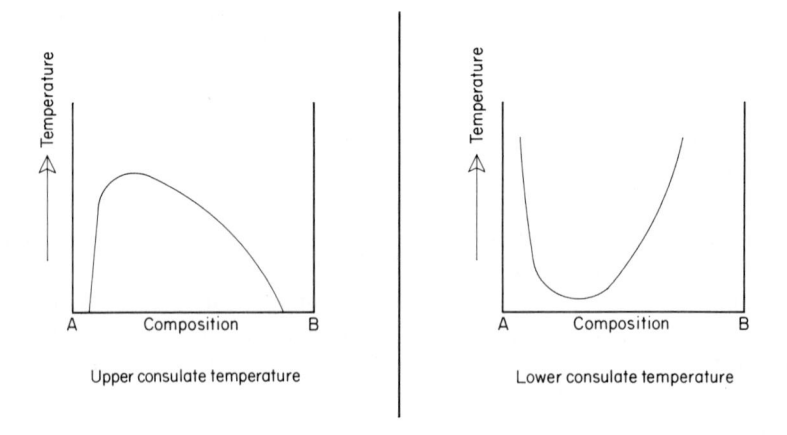

Fig. 11 Demixing as function of temperature

I. Liquid–Liquid Extraction (LLX) Procedure—An Integrated Approach 53

Similarly, disproportionation of the B^E phase by another additive F could be a viable approach, provided F exits from the system with the product to give $C^E + F$ as a separate phase, leaving B which is to be recycled, as shown in Fig. 13.

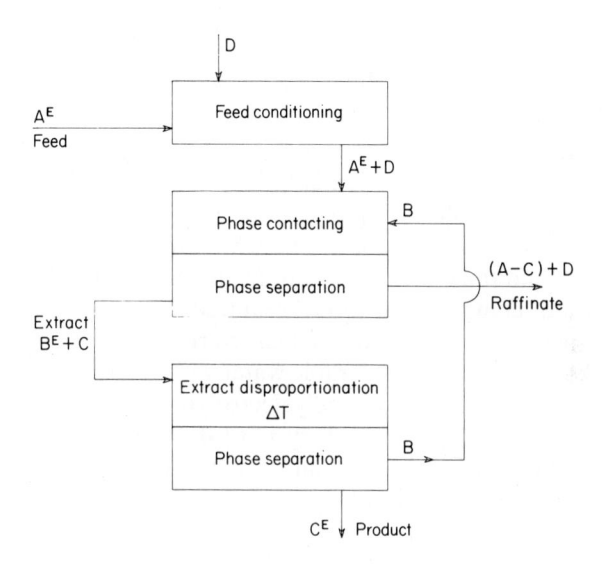

Fig. 12 Generation of two-phase systems by feed conditioning and temperature effect

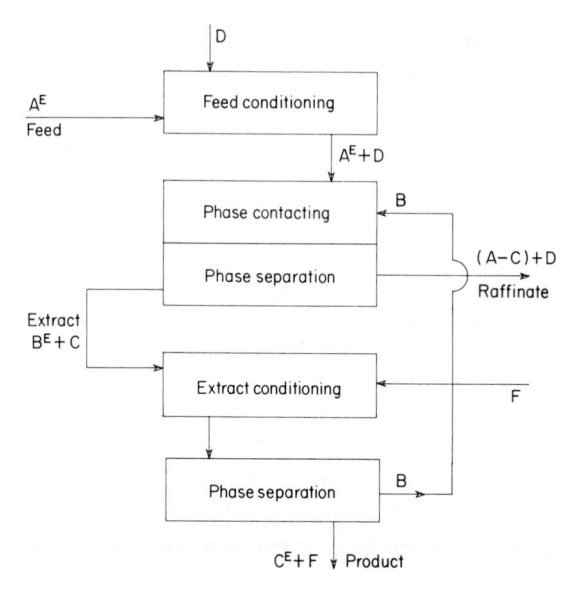

Fig. 13 Generation of two-phase systems by conditioning both feed and extract

When feed conditioning is required by adding a component (D), the extraction simulation can readily be countercurrent multiphase; similarly, when extract disproportionation requires adding a component F, a countercurrent stripping system can be envisaged if this is favorable. When a temperature change is used to produce phase disproportionation it may be sufficient to utilize a single stripping stage—a knowledge of phase compositions and distributions is necessary before this can be determined.

In every case the questions to be asked are how to generate the two-phase system(s) for extraction and stripping, such that the desired material [e.g. enzyme (E) produced in a fermentation] will distribute from feed (A) to phase (B) and from phase (B) to product (C) to liberate phase B for recycle. An understanding of the property of E which makes it favor B or C according to imposed conditions will help in optimizing the overall system.

This would seem to be the model required: the composition of B must be such that it has strongly aqueous characteristics; the polyelectrolyte itself may be able to take up considerable water without itself dissolving in water; alternatively a mixed solvent—polyelectrolyte + S—will be required.

The challenge for biotechnology is to devise systems for transfer and separation without losing the characteristics of LLX technology, that is, without sacrificing the concepts of continuous steady state processing.

The essential characteristics of LLX can be summarized as follows:

(i) two mutually immiscible liquid phases must be present
(ii) mass transfer occurs across the liquid–liquid boundary
(iii) possibility of devising systems with a high degree of selectivity and/or rejection
(iv) reversible transfer and steady state operation
(v) well developed technology
(vi) mechanism of transfer defined
(vii) desirable characteristics of a LLX system are:
　　　chemical stability
　　　cyclic, steady state operation and processing
　　　process and operating control
　　　technological feasibility
　　　favorable economics

For separation and recovery of biologically synthesized products, analogies can be drawn from other similar materials, or from other procedures for the same materials. The primary environment will be aqueous.

LLX in Conjunction with Fermentation

The economic use of micro-organisms for producing useful products from low cost substrates, may be "fermentation-limited" in some cases. This limitation may express itself in various ways. Thus in some fermentations the products may have a toxicity level for the organism hence limiting the concentration of product attainable; in other cases the product may cause high viscosity in the fermentation broth thus limiting the efficiency of phase contacting, gas/liquid and liquid/solid; yet another example is the case where the product is consumed by the organism, e.g. for organism maintenance, once the level of product becomes competitive with the level of residual substrate.

In chemical technology there are cases which can be regarded as parallel or analogous to the examples mentioned above, e.g. they are equilibrium controlled, or display an array of reactions in series, or show physical characteristics which interfere with operating efficiency. In cases of these types, LLX offers the possibility of removing the reaction product from the scene thus promoting conversion and increasing conversion yields.

In biotechnology, by analogy, the same advantages can be anticipated from two-liquid phase systems. Two operating modes can be considered: the one would permit fermentation to proceed in the presence of the second liquid phase, the other would require withdrawing fermentation broth continuously for liquid–liquid extraction and return to fermentation. In chemical technology, both procedures are applied, so that the liquid–liquid contacting/separation steps are well integrated into the overall operating schemes. Before this can be done in equal manner in biotechnology, it will be necessary to consider the specific requirements and constraints imposed by the "bio" nature of the operation.

Clearly the first aspect in all cases will be to select the second liquid phase to have minimum effect on the fermentation environment of the organism; this probably means that the second liquid phase should have very low solubility in the aqueous phase, it should not "wet" the organism (i.e. it should not displace water from the organism surface), it should not lower the surface tensions and interfacial tensions in the system causing emulsions etc., it should have a high distribution coefficient for the product, but not for other components in the substrate, so that it acts as a selective extractant; there should be no difficulty to release and recover the product from the extract phase so as to return the extractant to the fermentation system. If the distribution coefficient is such that full substrate depletion can be attained before extract treatment need be undertaken the extractant would be returned to the next fermentation; if a type of steady state continuous fermentation can be envisaged, then the extraction too may require to operate in conjunction with fermentation with some sort of steady state.

Both the operating modes postulated above for biotechnology differ from the well-established procedures of using LLX as a recovery/purification method *after* fermentation, in other words, they differ from the systems used (e.g. for penicillin recovery or for citric acid recovery or in enzyme recovery using biphase aqueous systems), as all these procedures are *post fermentation* steps and not integrated with fermentation.

Flowschemes can be postulated for the integrated approach and the general constraints of such systems anticipated. The background studies required in order to develop these approaches are not related in the primary instance to the mass transfer aspects since these can be examined outside the framework of fermentation. For studying the interactive aspects it will be necessary to call on a broad array of disciplines in order to anticipate organism/extractant interactions and possibly to avoid them, so as to select suitable extractant candidates for specific cases.

Envisaging New Process Applications for LLX

It may seem odd that notwithstanding intensive R & D work and much progress in LLX technology in the past two to three decades, the actual examples of the technology in operating industry remains small. In organic chemistry and in pharmaceutical chemistry, SX by partitioning is an accepted procedure, along with crystallization and distillation, for separation/purification/recovery, but this type of SX is not what we are looking at in the present context. Rather we are thinking of solution chemistry with strong interaction in solution, where the solvent composition can be tailored for the particular purpose to promote transfer for achieving a particular aim.

The type of SX or LL distribution which was commonly applied in organic processing was indeed essentially a "unit operation" but LLX with strong solution interactions has probably more in common with unit processes than with unit operations. The moment one accepts this postulate, the situation which appears odd becomes instead clear and obvious. Thus, for example, if one were reviewing distillation, it is unthinkable that the materials being distilled would form the basis of classification, but this has been the case with LLX. In fact it is becoming more so as developments are "in depth" developments of specific cases. Now, when one looks at unit processes, the classification there is logical and limited; so too if one utilizes "mode of transfer" as analogous to the unit process, the classification of LLX becomes logical and limited. This, in fact, was the classification postulated initially for LLX, being based on mode of transfer:

— Partitioning—"inert solvent".
— Compound formation, e.g. with chelating agents or acidic reagents.
— Solvation, e.g. with C-O or P-O containing reagents.
— Ion-pair formation, e.g. with amines or acids.

This generalizes the concepts and thus serves as a means of transferring "learning" or experience from one area to another.

The shapes of the equilibrium curves that can be anticipated on the basis of mode of transfer are of three types, as shown in Fig. 14; the first is stoichiometrically controlled, as in Fig. 14(a), and applies when molecular ratios are limiting as in compound formation, or in ion-pair formation; the second is controlled by concentration as with solvating reagents, shown in Fig. 14(b), where the extracting solvent is in competition with the initial solvent; the third case, given in Fig. 14(c), is simply a function of relative solubility of the solute in the competing solvent, without strong interactions.

In looking at LLX applications the classifications by these three types brings together cases which otherwise appear to be absolutely unrelated. Hence know-how which is not being transferred and is essentially being lost as far as new applications are concerned, could be used as analogy for evaluating LLX as a separation technology for new, unexplored applications.

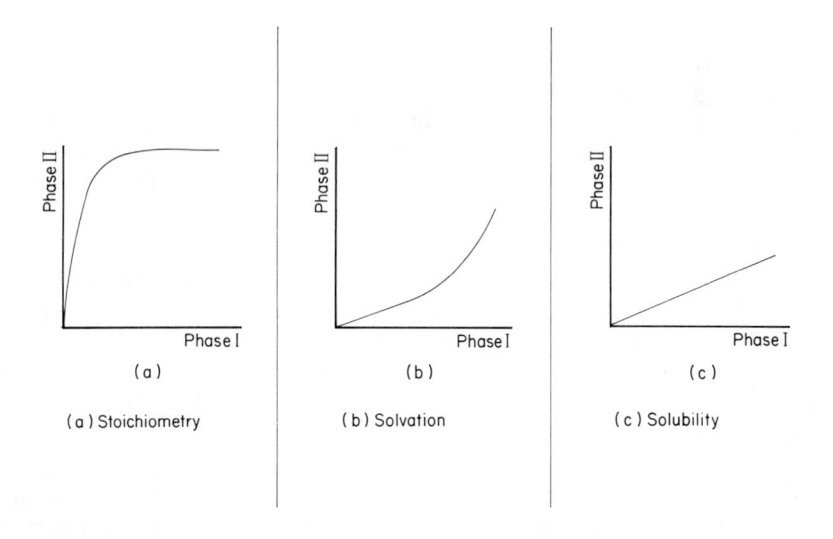

Fig. 14 (a), (b), (c) Equilibrium curves

If one were looking for new applications for LLX one could go by different routes. One route would be to use some scale of production volume or value as a yardstick for selection of likely cases, review the respective production flowscheme for these cases and classify them according to the basic modes of recovery, so, for example, in some cases insoluble salt production is the central separation step whether by precipitating acid or base values, or by depression of solubility say by pH change, or by common ion effect or by temperature change, by double salt formation and so on. In other cases, reaction may be the mode of separation, e.g. in metathetic reactions, ion exchange, neutralization etc. A second route would be to classify materials by types and then to look at production flowschemes, and especially the main separation steps. If neither of these schemes seems realistic, possibly what one needs is a summary of what LLX can do in generic terms, and what this would be based on. Then those who do separations would be able to ask "what can LLX do for our case?" Actually LLX seems to be the only separation which has not been generalized, but has been kept to specifics, almost always classifying by what is being separated and not by the nature of the step which permits attaining separation. Probably the successful approach would entail a combination of specific procedures used for identification and analysis, with freedom of choice of components to constitute the LLX system. Thus when one asks "can one recover A by LLX?", the counter questions are "how do you recover A now, how do you analyse for A, what are the materials B, C, etc., that A must be separated from? How do you separate A from B, C, etc., now, how do you evaluate your success now?" In retrospect, in almost all cases requiring separation of A from an ambience, it has been possible to postulate a LLX system for doing this; not always is the system technologically viable, but that is the second simpler stage of the problem, since technology changes and develops, hence viability depends on the times. The main requirement first is "feasibility in principle" and that demands much imagination in regard to adapting the "known" and the "possible" so that together these add up to the "probable".

Based on the answers to the above questions and more, one can attempt to define the nature of the second liquid phase which will constitute a part of the LLX system. Thus one should be able to judge how similar or how different from Liquid I should Liquid II be. One should be able to judge what type of transfer is to be utilized, in other words what one would select for promoting the transfer.

Let us take the following case: the question asked was "can one devise a LLX system for separating A, an amino sugar produced by fermentation, to replace a complex adsorption/ion exchange/chromatography system?" The answer was "probably, because in principle what can be done by solid ion exchange can be done also by using a liquid ion exchanger." The type of ion exchanger being used was strongly acidic since the amino-sugar was very weakly basic, hence a sulfonic acid was chosen for the separation—dinonylnaphthalenesulfonic acid. The diluent had to be selected taking into account that the more polar the diluent, the more basic the reagent becomes, hence here a non-polar diluent would be preferable. The feed contained a significant quantity of cations which would be in competition with the ion exchanger, hence pH was a very important parameter in controlling the ion exchange separation. It was found indeed that LLX, by virtue of its versatility in regard to the composition of the second liquid phase, could indeed give a positive answer to separation feasibility, although not necessarily to quality. Naturally viability and desirability required a more critical evaluation.

What this boils down to is that envisaging new process applications of LLX needs the cooperation of those who know the current processes, understand how and why they are operable, and those who know LLX and what makes it work.

The parties who are actively involved with LLX are:

(a) those doing basic research
(b) those doing equipment development, equipment vendors
(c) reagent suppliers, reagent developers
(d) those using a specific LLX process

Who is likely to be interested in developing new processes or new applications utilizing LLX? Categories (b) and (c) would be interested provided the return on the R & D involved will pay for the R & D invested. This means that (b) and (c) are not seeking new processes *per se*. Category (d) has interest in expansion in the sense that one feels at home with LLX and is therefore not hesitant to expand its use. Category (a) will usually have an intensive approach, probably not end-use oriented. However, if (a) has an extensive approach and is end-use oriented, then novelty should be the essence of the research (as opposed to understanding only the accepted), hence category (a) becomes a starting point for new processes and new systems.

Indeed it is correct that new angles, new approaches, and postulated new areas of application will come from category (a), but implementation will not, since often these new areas are not the ones with pressing reasons for taking up novel approaches and developing them into innovative processes.

The ideal would be for the user to publicize his requirements and his dilemmas so that the inventive researcher can propose approaches to novel routes and solutions. Unfortunately this rarely happens since no user will publicize his problems or his innovative plans. This means that innovative processes using LLX can usually materialize only where the user has a sufficiently high level of "in-house" expertise and inventiveness at his disposal. This of course is by no means a fact in all situations.

A different approach is to promote the scope of LLX to such a degree that it can become actively a part of anyone's professional expertise. In other words, every graduate "applied" scientist/engineer should know that LLX is a versatile tool for consideration in all cases where the presence of two phases can be put to positive use. The positive use may be separation, recovery, keeping separate, bringing together, whatever promotes the procedure being sought.

This starts therefore as an educational problem. Presently LLX is usually taught as a simplistic distribution operation between two liquids of differing polarity; the broadness of the scope and the broadness of the choice of system is usually not presented adequately. Perhaps it is this very broadness that makes it difficult to present a limited package.

Hydrometallurgy did a great service to LLX and that is probably where the action is still today. Similarity and analogy play no small part in this case, so one or two great success stories have been enough to fire the imagination and to keep the fires burning. There are not that many other general areas where similarity and analogy apply to such a degree—biotechnology is one that comes to mind at this point; the need for separation and recovery from dilute solutions will become increasingly pressing as renewable starting materials take on an increasing role. It is here, therefore, that similarity and analogy can be expected to play a part in expanding the application of LLX provided that one major success story can be described. There is a fallacy in the thinking that in LLX the separation problem is simply transferred from one solvent to another since eventually the product must be separated from the second solvent, e.g. by distillation. Yet in many cases this is absolutely not so. Transfer may be parameter controlled, thus in the large volume extraction plants distillation for separation is rarely used; the important point, therefore, is to select a transfer system which will permit some parameter to control direction of transfer. A common ion, pH, temperature all have been used and have been publicized to a limited degree, so these cases could serve as analogy. The big difference between the widespread publicity which has accompanied the Cu^{2+} success story, and the limited publicity for the KNO_3 success story, is that in the former case publicity was desirable for the reagent supplier who had been largely instrumental for the success, while in the second case publicity was thought to endanger marketing control hence the producing company preferred to keep the essential know-how unpublicized. This latter approach is common in industry even though it does defeat the wide purpose of keeping research in touch with the real problems.

If, therefore, it is not likely that research will always be able to attack the real problems that need to be attacked if implementation is to follow on, perhaps a similar purpose could be served if the generalized lessons learned from successful cases could be clearly stated, so as to serve as analogy models for other cases.

This requires identification of the characteristics of the system involved. Why does the presence of the two phases promote separation in the desired direction, what are the characteristics of the second liquid phase in the system and the mode of transfer, to what extent is this specific for the case in point, and whether it can be expanded to encompass other cases—of similar type or of different type? Would a different type of second phase be able to perform similarly, on the same basis or on a different postulated basis, and what are the deficiencies of the present system? Would the postulated system obviate these deficiencies; what the expected deficiencies in the second case would be; whether the second phase, used successfully here, could be applied to another case of similar, or of different character? How wide a range of differences can be spanned by a chosen system; what were the special technological and economic aspects that encouraged success in the reference case, can these be applied in a different case, are they related to the second phase chosen, or to the separation type itself?

Many such questions need to be asked—organized answers are likely to promote the application of LLX to other cases, similar or different.

Systematization of Solvent Systems

In phase transfer catalysis, ammonium and phosphonium salts, containing sufficient C atoms to make them sufficiently lipophilic, are both accepted. Thus, for example, tetra-n-butyl phosphonium chloride is proposed as a stable phase transfer catalyst. It is doubtful that phase transfer catalysis would normally be used as a source for analogies for LLX, even when ion-pair formation or ion exchange is the basis of both. Yet phosphonium salts and phosphonates along with sulfonic salts and sulfonates actually present themselves as the basis for a scale of bases and acids which then can be applied to promote narrow separations.

In regard to solvating solvents, the scale of "donicity" and donor numbers provides an interesting approach, but again this is not common jargon among LLX chemists. At most, dielectric constant will be a property that is used for comparisons, usually, however, relating to the solvent or diluent component and not to the whole system of components in the non-aqueous phase.

This lack of systematization seems to apply also to the active reagent in a solvent system. An increasing number of individual materials are being tested, but one does not get the impression that any systematic scheme is being followed in so doing.

Materials to be extracted *can* be graded in some sort of order, e.g. cations can be ordered according to valence and base strength. It would be convenient if one could say that if for M one uses X as extractant, therefore for P one could use either Y or Z. In other words, it would help if one could select the reagent according to some scale commensurate with the cation scale, by following known analogies. However, this seems far from reality at the moment.

There is another interesting aspect of solvents. In LLX all one needs are two phases, provided of course that the characteristics of the two phases are sufficiently different in relation to some property of the solutes to permit expectation of separation. In biotechnology an interesting advance has been made which conceptually should be considered for other cases too—this is that when two polymeric materials, both water miscible, but of different molecular weights are mixed in a three-component system, i.e. R_1, R_2, H_2O, two liquid phases will be formed, and certain materials will distribute between them. Apart from this case, and the case of phase generation by temperature change, there is also the case of "third" phase formation due to the change in polarity by virtue of the extraction itself. Yet this concept of creating two phases as a result of the system itself has hardly been exploited in separation technology. This approach would change the attitude to the standard type of flowsheet, which is a two-box concept with auxiliaries, but at this point it is not clear what the general format of such a flowsheet would be, or even whether there is place for generalization, *a priori*.

The concept of "salting out" or "salting in" is old, and is used continually in organic chemistry. In LLX to some extent it has lost its significance. However, when one looks at equilibrium constants and the nature of the various species involved, one really needs to look in a much broader way, if one wants to use such facts to help in designing or selecting LLX systems.

Possibly, too, differences in coordination numbers of materials, steric effects etc. are not nearly sufficiently exploited. If one takes a particular requirement, then one should be able to say what the expected effect of change in structure will be, and then to go on from there.

An example of such a modification of an extractant was the introduction of a halogen atom on the α-carbon of aliphatic carboxylic acids like lauric acid to give, for example, α-bromolauric acid, a stronger acid than the parent acid. This then permitted cation exchange at lower pH.

Modification of amines also leads to a scale of basicity, thus adapting the base to the strength of the acid with which it is to interact. Proper selection or modification of the amine will lead also to a scale of sensitivity to the effect of temperature on equilibrium constants for ion-pair formation.

It is not always necessary to make a change in the structure of the reagent itself, since in a multi-component system in which the second liquid phase consists of reagent, diluent and modifier, solvating effects can be utilized to change the relative scale of the extractants. This applies equally well with acidic or basic extractants, the interaction being between reagent and modifier.

In specific areas, where reagent manufacturers control the operation in the sense that they fit the reagent to the problem in hand, the manufacturer has very considerable in-house expertise; however, this expertise is generally not available in the public domain. There is, of course, no reason why an analysis of a problem should not be done by a group with sufficient joint expertise to be able to identify the requirements of each specific case and to specify the appropriate action for choice or modification of the reagent or of the liquid–liquid system as a whole.

In order for the latter approach to be viable, a strongly interactive inter-disciplinary team is required. In addition it will be vital to dissect each problem carefully so that required separations be clearly specified, also that examples of types of reagents which will permit such separations be indicated and that adverse aspects and problematic areas be defined, so that all in all the ideal or desirable reagent can be postulated.

This does not mean that individuals, working alone, have no place in the LLX research; on the contrary, the more there are in depth studies on vital aspects, the easier will it be for a multi-disciplinary group to identify the factors that are relevant to a specific case, so that interactive projects covering these factors will follow naturally when the development of a process is being considered. At the same time this should lead to systematization of solvent systems by the group identifying the interactions of systems and the requirements of separations.

Technological Aspects

Irrespective of the system or its purpose, liquid–liquid extraction by definition comprises a number of technological aspects which are inevitable, being inherent in the unit operation itself. Thus at least two liquid phases must be present; these are to be brought into contact so that material can transfer from one phase to the other, and the phases are then to be separated. Essentially there are two kinetic aspects involved, the rate of mass transfer as a function of the extent of contact, and the rate of coalescence. The approach should be to provide sufficient contact for good transfer kinetics, but not too much contact, so as not to influence adversely the rate of coalescence and phase separation.

The subjects of mass transfer as a function of dispersion, and of separation as a function of coalescence and physico-chemical parameters have been studied in great detail and there are a large number of good publications on the subjects. Anyone who is active in LLX process development needs *background* in these topics, but often only so as to avoid coming up with flowschemes or procedures which negate good technological practice. It is always better to anticipate technological constraints which may have bearing on implementation, particularly if such contraints have only a minor relationship to process and can easily be taken into account from the beginning. Alternatively, aspects which may have a bearing on technological implementation should be stated, together with those process aspects that are open to change, in order to accommodate technological considerations.

Let us indicate a case in point. Let us assume we have a system which tends to form a third liquid phase which would cause difficulties in the separation of phases in certain types of equipment. This problem can be obviated by adding a small amount of a polar solvent as modifier, e.g. an aliphatic alcohol. However, an alcohol is an amphiprotic solvent which may have a strong interaction with the extractant, e.g. in an acid–base system, and also may itself distribute between the two phases. If phase separation considerations require prevention of the formation of a third phase, it may be necessary early in the PD to review the distribution of components as a function of the third-phase preventor, so as to fit the case to the constraints.

Alternatively a different approach may be to modify the equipment so as to be capable of handling three phases without changing the physico-chemical system in any way. It must be remembered that an additional phase eliminates a degree of freedom, thus helping to make a system self-controlling. Another point is that the more complex the solvent, i.e. the larger the number of components which go into its constitution, the more the chance of unmonitored and even unexpected variations in solvent composition. Thus, when using a solvent, composed of an extractant, a diluent and a modifier, the solubility losses to the aqueous phases with which the solvent comes in contact may differ for each of the components, so that one is likely to experience solvent composition drift, unless there is a change in the solvent recovery step. These observations indicate the indirect bearing which the system may have on technological aspects and which must be realized in order to be taken care of.

Process control should be distinguished from operation control. In a LLX system the process is determined essentially by distributions and by mass balances, the former expressing itself in concentrations, the latter in feed and product rates. The process is accordingly controlled by flows and by compositions. Operation control is aimed at keeping the *status quo*, by not allowing it to drift from steady state. In a LLX system, apart from flow control, aspects of phase contact, i.e. dispersion, and phase separations, i.e. coalescence, are of overriding importance, since they control the efficiency of mass transfer, which is what the whole operation aims at.

In the event of drift of the component of interest, say by the rising concentration in the raffinate, it is necessary to determine whether the process control or operation control is at fault. The central process control points will relate to concentrations and phase ratios, i.e. material balances.

In designing the process control during the PD study, the sensitive control locations should be identified. Thus in a battery of mixer settlers the raffinate itself may not be the best test point; instead, if the distribution is strongly concentration dependent, the inflection point will be the best control point in the battery. Another example, if it is known that there are minor components in the system which may accumulate in the solvent and thus affect its quality and efficiency, this aspect will need to be controlled. In certain cases the effect of accumulation may itself be used as the basis of the mode of control; thus if *coalescence* is adversely affected by accumulation, a standard settling test may be the best way of recognizing accumulation. Knowing that minor components are building up is one thing, but their elimination is another thing altogether. This brings us to two aspects of operation control, the one aims at ensuring an open-ended system, so that what comes in can get out, the other is an approach to solvent clean-up. The means selected for solvent clean-up are very much dependent upon the chemical characteristics of the solvent system and of the impurities. Thus distillation is an obvious mode of clean-up; similarly washing under selected conditions, alkaline or acid, to cause impurities to leave the solvent by precipitation, back extraction, complexation, etc. The aim here would be to keep the impurity level low enough so that the rate of coalescence is acceptable. Coalescence is not only sensitive to quality of solvent but also to phase ratio. In some cases recycle of equilibrated phase so as to change the ratio may be an efficient way of controlling coalescence, without interfering with distributions and equilibrium.

There is another aspect of impurity accumulation and that is *precipitation* inside the system, leading often to solvent wetted solids collecting at the interface, and interfering with proper operation of the equipment. The source and cause of such solids needs to be identified so that the problem can be handled if not eliminated.

Deterioration of solvent quality may be reflected in the mass transfer itself. Thus when accumulation is permanently binding a part of the active transfer reagent this will express itself in summation as a change in the equilibrium distribution curve or in the capacity of the solvent, i.e. the degree of loading that can be attained. Here too the change in the aspect being affected then becomes the control for the accumulation.

Economics/Engineering of LLX Operation

LLX equipment must perform a number of operations: it is required to move liquids (often in opposing directions) to mix liquids, or let us say to cause them to be in contact so that mass transfer can take place, then to promote separation between the liquids after mass transfer, and to move the liquids onwards. There are essentially only three formats of LLX equipment—mixer settlers, columns, centrifugal units. Others, such as rotating film contactors, have been proposed, but have not reached the forefront in equipment types.

Equipment selection has a strongly interactive dependency, being controlled by the liquid–liquid system characteristics as well as by the equipment characteristics in relation to the liquid–liquid system. There is also a strong interactive dependency with the plant as a whole.

Purpose and desirable characteristics of LLX operations are summarized in Table 14 as a basis for approaching economic and engineering aspects of the operation.

We can assume that most LLX operations will be single or multi-stage, and that most are equilibrium controlled. There are certain separations that are kinetics controlled, for example, when the rates of formation of the complexes between individual extractable components and extractant are sufficiently different to permit separation of the fast forming complex from the more slow forming one(s). However, usually the rates of transfer and

TABLE 14. LLX.

Purpose	Separation, recovery, purification, concentration
Mandatory	Two-liquid phases
	Mass transfer across interface
Desirable characteristics	— Three phases involved
	Phase A—feed phase, given
	Phase B—transfer phase, generated or selected
	Phase C—the recovery phase, generated or selected
	— Cyclic steady state operation
	Transfer A → B; B → C
	B returns to A
	— System: Chemical stability
	Acceptable selectivity and/or rejection
	Control—process and operation
	Technological viability
	Economics favorable
	— Transfer: Reversible, i.e. into B, out of B
	Water rejected, i.e. from or with A

complex formation are rapid and the reactions are reversible. Equipment selection must therefore relate to both these aspects. Response to an imposed change should be rapid, while response to an inadvertent drift should possibly be slow enough to prevent the whole LLX system from going out of steady state before correction can be applied. An important aspect in equipment selection is the degree of certainty that relates to scale-up. The approach here will differ, depending upon whether the equipment is to be an "in house" design and scale-up, or is to be supplied by a vendor under guarantees. It is clear, of course, that guarantees are of little use if the scale-up know-how of the vendor is not adequate. In certain cases there may not be sufficient information about secondary accumulative effects due to the characteristics of the system, hence these may not be given enough weight in scale-up.

The economics of solvent extraction systems can be separated into the LLX operation itself and its interaction with the other process steps which surround the LLX operation.

Mixer settlers require a large inventory of solvent, columns require an intermediate volume, while centrifugal extractors have minimum holdup. Mixer settlers can be shut down and restarted without losing the concentration gradient or profile across a multiple-stage battery; a column on shutdown loses the profile, so that contents must be reprocessed on restarting. Centrifugal contactors do not have such problems and can be started and shut down as desired, since holdup is so small. On the other hand, any change in feed to a centrifugal contactor will be reflected immediately in the streams leaving the contactor, whereas with a mixer settler it will take some time to observe a real change in the streams leaving the unit.

The coalescence characteristics of a system as a function of the dispersion type may have a direct bearing upon equipment selection. There is often a very large difference in coalescence rates between an "oil-in-water" and a "water-in-oil" dispersion. Under gravity separation it may be necessary, therefore, to control dispersion type so as to keep equipment to reasonable dimensions; while the mixing may have a marked effect on the dispersion generated, the only independent system parameter controlling dispersion type is phase ratio. If the "natural" phase ratio does not favor the desired dispersion type, it is possible to arrange for internal recycle of one phase in order to attain a suitable phase ratio. If the equipment type is sensitive to total liquid throughput, this internal recycle may have a detrimental effect.

In mixing equipment, whether this be the mixer of a mixer settler or the mixing zone of a mechanically agitated column, it is possible to select the dispersion type initially and to start in this mode. At operation, within the capacity limits of the equipment, this dispersion type will be maintained, but a flow disturbance or increase may cause inversion of the dispersion with consequent problems in the coalescence zone of the equipment.

In the case of centrifugal contactors, the dispersion type can be controlled by operating conditions, e.g. by the pressure differential, hence recycle to correct the phase ratio is not required. Also, both dispersion types will be present in a centrifugal contactor.

Even with the desired phase dispersion type, the viscosity of the system may have a bad effect on coalescence rates. If so, it may be advisable to increase the temperature by some degrees in order to cause the viscosity to drop. Naturally the assumption is that a somewhat higher temperature will not have an adverse effect on solvent stability nor on distribution coefficients in the specific case. Alternatively a diluent may be used in order to reduce viscosity (if this does in fact do so).

Equipment Characterization/Basic Aspects in Relation to Equipment Selection

The operation of LLX entails two steps, mass transfer and phase separation. Normally when bulk volumes are flowing, dispersion of one phase in the other is the mode of attaining sufficient transfer area for mass transfer, requiring therefore recoalescence of droplets to permit phase separation to take place. Liquid membranes do not necessarily require these steps, so they are excluded from the present discussion.

For design there are various accepted simple procedures for determining mass transfer rates in a practical manner; alternatively, residence time and energy input may be the reference for scale-up, for a selected equipment type. Recently a number of three-component reference systems have been selected so as to permit valid comparison between equipment types. This is a great help for excluding unsuitable equipment, but it does not eliminate the necessity to do positive tests with the actual LLX system being considered, since presumably equipment types can be adapted to particular cases. In other words it is not sufficient simply to obtain a model unit from a vendor, to operate it according to standard procedures, and to scale a full unit from there. In the model unit some aspect will be limiting, depending on the system being tested; this can result, for example, in a test column giving say two theoretical stages at some throughput for one pair of liquids, but say six stages at the same throughput for another pair of liquids of the same family. Where it is possible to describe the system in the two cases, especially by putting numbers on relevant characteristics, and where it is known how these characteristics relate to the intimate design of the equipment type, it may well be that a small modification will permit adaptation of the same type of equipment for both pairs.

Similarly, it may be misleading to use data obtained in one type of contacting equipment as if these data were basic and relevant for a completely different type of equipment. Fundamentally, this may indeed be so, or perhaps one should say in the limiting case this will be correct, but then the limiting case must be used for the cross-over from one type to another. Thus, for example, residence time in a mixer has no relation directly to contact time in a centrifugal contactor. Obviously coalescence time in the two cases too cannot be related because of the centrifugal force in the latter equipment which determines phase separation.

From the above it follows that for a novel application, or even for a variation of an accepted application the choice of equipment cannot simply be based on a practical test program. Two things must interlock: namely the process demands and the equipment attributes. The process demands will relate to what is the stated aim and what are the aspects that accompany the steps leading to the aim. A simplistic example would say that the aim is to obtain as concentrated an extract as possible according to the distribution coefficient; the accompanying fact may, for example, be that such an extract will be very viscous, hence it is not the feed streams to the contacting equipment that will be determining, but the exiting streams. The equipment attributes may be such that viscosity plays a very limited part in influencing capacity, or the equipment may have a viscosity cut-off level.

Probably a workable approach would be to list equipment attributes and limits and to balance these against process demands.

Environmental Interactions

LLX implies the presence of at least one liquid phase which contains either a single non-aqueous material or several such materials. This is so, whether or not the other liquid phase is itself designated as aqueous or is also essentially non-aqueous. Thus in separation of aromatics from aliphatics by LLX, neither phase is aqueous, but both examples satisfy the definition above. The environmental interaction derives essentially from the non-aqueous components of the system and will relate first of all to the properties of these components, such as volatility, solubility, flash point, etc. The implications of these properties are straightforward and must be taken into account in equipment selection and plant design. Other interactions are more complex and will be dependent on the system itself. Thus, for example, amine solubility in water may be extremely low whereas amine salt solubility may be considerably higher. Similarly, an organic acid may have very low solubility in water while its salts may have real solubility, or may cause emulsions to form. The presence of modifiers in a system which can cause azeotrope formation, for example, or the existence of an azeotrope with water itself can change volatilities very much when compared to the separate materials. Similarly modifiers can change mutual solubilities, and this must be recognized if the interactions of system and environment are to be well anticipated.

The aspect of environmental interaction in relation to disposal of some material exiting from a LLX system is very important, and may necessitate positive steps to reduce the level of potential contamination to the required limits. Yet another aspect of importance is the potential attack on materials of construction used outside the LLX system.

In the framework of the LLX process, it is clear that all these aspects must be taken into account right from the start. Another aspect that has a different implication relates to produce quality and specifications. This is an environmental interaction in the broadest sense. Thus, as an example, any carryover of solvent phase from LLX to electrowinning in say a copper-winning plant may have far-reaching implications. Another example would be, for example, the neutralization for waste disposal of an acid or base stream produced by LLX, whereby dissolved solvent components of base or acid character would be liberated as active contaminant.

Peripheral Operations/Interactions with LLX Step

Since LLX is a separation tool, it will rarely constitute the process as a whole, but will usually be interactive with other operations. One rare exception is the metathetic reaction of KCl and HNO_3 to give KNO_3 and HCl. Here the whole process is solvent extraction dependent; there is no feed preparation, and apart from drying, there is no post processing (see Fig. 3.3). Usually, the situation is quite different. Take two cases: hydrometallurgy and fermentation.

In hydrometallurgy there will usually be a leaching or dissolution step before the LLX, and a metal or product recovery post-LLX. If the liquid raffinate from the LLX step is returned to leaching, then LLX interacts twice; also the LLX interacts with recovery, possibly twice. This is shown in Fig. 15(a).

In fermentation, the final broth will contain a whole gamut of materials besides the main product, deriving from the initial substrate or from the fermentation itself. Recycle is rarely practised; recycle from LLX upstream would require particularly careful consideration because of the danger to fermentation of carry back of solvent components. The interaction therefore would be mainly a forward interaction between fermentation and LLX, while the LLX recovery interaction may go in both directions, as can be seen in Fig. 15(b).

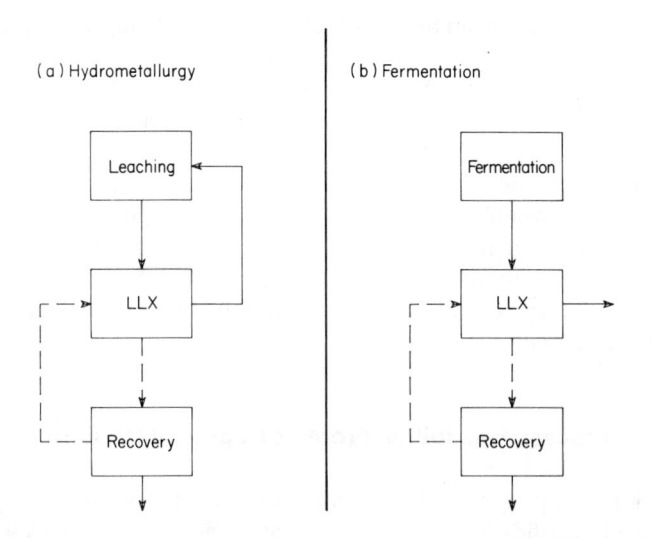

I. Liquid–Liquid Extraction (LLX) Procedure—An Integrated Approach

Fig. 15 Interaction between LLX and other steps (a) Hydrometallurgy; (b) Fermentation

Cost evaluation in all cases will depend on the added value of the product, as a function of the separation achieved, versus other possible separations. If there is virtually no other separation procedure applicable, the process stands or falls based on the efficiency of the LLX step and its interaction with the other process steps. When there are various procedures for separation of product, then LLX can only be weighed in each case against the common practice.

A strong interactive effect in all cases will be that related to solvent losses due to solubility, or to entrainment, or to secondary carry over of some sort. This interaction has several important aspects; firstly process-wise, when there is recycle as mentioned previously or secondary use of raffinate or drag, secondly environment-wise when there is a question of disposal of raffinates, thirdly product quality aspects, fourthly variable costs due to solvent make-up.

In general, it is desirable to have the feed to LLX sparkling clear, since even small quantities of finely divided solids can cause solvent losses by adsorption; this solvent wetted particulate material then causes further problems by accumulating at the liquid–liquid interface, interfering with phase separation.

Surface active materials present in the aqueous feed may cause a sharp change in interfacial tension in the LLX system, completely changing the dispersion/coalescence characteristics and hence possibly the scale-up factors in relation to equipment and even to equipment selection.

The water balance is an aspect of LLX which is strongly interactive with process steps, before and after the LLX. First of all this is almost obvious, since distributions are concentration controlled. Thus if a solvent picks up water in stripping and returns this water to the aqueous phase in extraction, there are two negative effects: one is the dilution of the solute in extraction which will be detrimental to extraction, at best requiring additional stages, the other applies when raffinate is returned for recharging since there will be a steady increase in its volume. If the water transfer takes place at an invariant point, again at best the negative aspect will be the steady increase in volume of the aqueous phase. This will require bleed of the aqueous phase and with it whatever is dissolved in it.

Process Feasibility, Proof vs Degree of Certainty

Verification of a proposed flowscheme in LLX is fairly simple. What is far less simple is verification of a LLX operation as this interlocks with other steps of a process as a whole. Process feasibility, which includes economic viability in the final instance, is essential, hence the degree of validity of the proof is a topic for consideration. In an integrated system, if there is sufficient analogy from which to draw, or if experience has shown what are the likely problem points in such a system, acceptable tests can probably be devised to strengthen the degree of certainty in proving process feasibility. It is when there is only a very limited background on which to base analogy or anticipation that the degree of certainty may become problematic.

Within the limits of available equilibrium data, with a reasonable scope of relevant testing, one can attain full certainty of the basic flowscheme. Aspects which are more difficult to quantify are, for example, solvent losses and loss of values in raffinate and waste streams. Usually, one differentiates between chemical losses and physical losses; the former are process losses, the latter operational losses. Physical losses are partly connected directly with equipment type (presumably one can get equipment vendors to guarantee the range for entrainment and venting losses); other operational losses are more difficult to foresee. Design of plant should take all of this into account ensuring post-separation and recovery of solvent as far as possible.

Chemical losses are a different story. First an analysis can be made of probable or possible routes of solvent reactions which can lead to losses, then a test program can be initiated to test the extent of such losses within the framework of the extremes of process conditions. This is reasonably straightforward, and reliable to the degree that the assumptions are reliable. The real problem comes of course from unanticipated reactions, since these will come to light *post factum*. This applies also to all unanticipated interactions which reflect eventually on process feasibility. Questions are how far should one go in trying to attain a foolproof system; when should one take a conscious risk, for example by establishing limits that can be tolerated, and assuming that one will be at the extreme of the limit; what does one stand to lose if one waits for a higher degree of certainty, and what risk is one taking if one does not wait? Coming on line early with a new process can compensate for a fair amount of uncertainty and risk.

LLX Detriments

Probably one of the most significant detriments of LLX is the large volumes to be contacted and separated. This is due to the kinetics of mass transfer and of phase separation, but also to the transfer mode itself which may often be concentration controlled or dependent on molar ratios of solute to extractant, and to the relative molecular weights.

In order to diminish this detriment, various aspects can be attacked:

1. Enhance the mass transfer by increasing the distribution coefficient* over the whole range of extraction. It is important that this increase apply over the whole range, since the lowest distribution coefficient will control the volume ratio of the two liquids.
2. Decrease the distribution coefficient* in stripping so that the final product volume will be appreciably less than the feed volume.
3. Enhance the rate of mass transfer if this is the controlling kinetic factor, i.e. if mass transfer rate and not coalescence rate is the determining factor for residence time. In order to approach this aspect with reasonable efficiency, it is necessary to have some picture of the rate controlling step of the mass transfer.
4. Promote coalescence and phase separation by appropriate steps. This will clearly be in part dependent on the mechanical transfer equipment being used.

All the above aspects have been dealt with successfully individually in specific cases; now an overall analysis of the problems aimed at the execution aspects of the technology is required.

Let us consider enhancement of mass transfer to increase the distribution coefficient. Combination of reagents which act "synergistically" has been studied in specific cases. This approach is particularly successful when it increases selectivity and sensitivity to process parameters. The problem of a

*Distribution coefficient: $K_D = \dfrac{\text{organic concentration}}{\text{aqueous concentration}}$.

high volume ratio often remains, however, because of the high molecular weight of the extractant, which makes reagent concentration the determining factor for the volume of the extractant phase. This can be exemplified in the case of tertiary amines or quaternary ammonium reagents. With a molecular weight approaching 500, a 1 M solution will have a 50% weight concentration and this is probably the best that can be used. If the stoichiometry of the extracted species is 1:1, the extract will be 1 M in solute too. When low molecular weight materials are being extracted this means that a relatively dilute solvent solution is obtained on a weight basis, even though in some cases it may be considerably more concentrated than the feed. This is so if the distribution curve has the shape shown in Fig. 16.

If a high molecular weight material is being extracted this may mean that a 1 M reagent solution will be too concentrated for use.

Coupled transport, used in membrane transfer, and also in emulsion liquid membranes, has probably not been exploited to the full in liquid–liquid extraction, or at least not consciously, although it is exploited in metal extraction where, for example, the final product concentration is dependent on the coupled proton transfer and is essentially independent of the feed concentration.

Now let us consider the case of hydrated crystals or solvated crystals. In LLX, precipitation of an insoluble complex is avoided. Why? Even formation of a separate liquid phase containing a stoichiometric relationship of

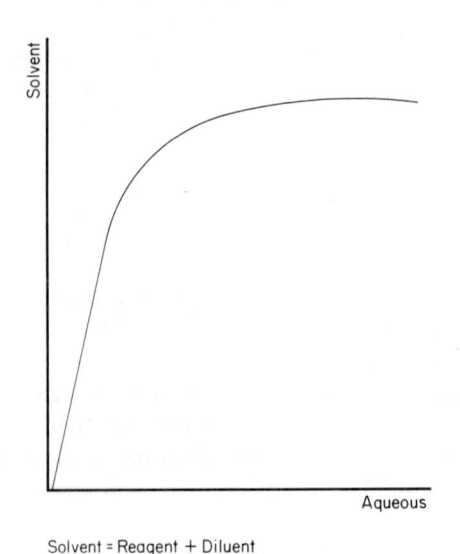

Fig. 16 Distribution controlled by stoichiometry

reagent to solute, with minimum diluent is avoided. Why? Certainly these will be more concentrated solute/reagent combinations than the diluent solution of the extracted species.

The reasons for avoiding phase changes derive mainly from equipment constraints and from lack of familiarity in the handling of such situations. Nevertheless, solid or high viscosity phases that collect at a LL interface, because they are solvent wetted or are not of sufficient density to report to the aqueous phase, could be the source of a high concentration extraction product, provided the formation of such materials by interaction of solute and extractant can be promoted. In other words, precipitation or phase separation by virtue of the extraction may be desirable phenomena. Indeed particular LLX processes, in which aqueous wetted crystals are produced by virtue of extraction, have been brought to large capacity industrial scale (KNO_3 process). In principle, there is no reason why solvent-wetted extractant–solute solids should not be equally useful technologically. In biotechnology where the use of biphase aqueous systems has been examined as separation systems, cell debris or solid organism from the fermentation step can be handled provided these report to one phase. In other systems where diluents and modifiers are used to overcome problems of limited solubility of complexes, it may well be that elimination of the additives so as to favor precipitation of extractant–solute complex may be a positive approach provided this complex reports to one phase. It is of course clear that techniques for handling such precipitated materials will probably have to be sought outside the normal competence of LLX. Thus the techniques and concepts of flotation, flocculation, sedimentation, electrophoretic precipitation etc. have hardly been considered in conjunction with LLX. These techniques have usually been considered for aqueous systems, but even in mineral processing non-aqueous systems such as heavy liquids have been studied. Factors which hindered implementation on such large scale would possibly not be detrimental on the smaller scale of specialized LLX processes.

In order for these concepts to be available for practical consideration it is necessary to set up a framework of general study of principles so that data are available to help those with real separation or recovery problems. Such a study must be an intimate interaction of chemist and chemical engineer so that some programme of limits can be set up. It can probably only be done as a sponsored project in the framework of a university or as a basic background project in the framework of the central research division of a large industrial company.

As an exercise, a framework for such a study project will be developed here:

Let us take as example ion-pair extraction where the ion-pair has limited solubility in the diluent. Two cases have been encountered, the one in which a third liquid phase is formed, the other where a solid phase separates. A study of such systems for useful application would need to concentrate on both the chemical and the physical aspects since there are two basic questions, one is how to promote the formation of the additional phase to best advantage, and the other how to separate the phases. A reasonable understanding of the shape of the phase diagram and of the limiting parameters which control the multiple phase equilibria would be necesary for assessing the usefulness of entering the multi-phase region. From limiting compositions a material balance can be derived; also it can be determined whether a pseudo-invariant system can be devised to give a self-controlling system. When the additional phase is liquid, one can ensure that there will not be entrainment of either of the other two phases, so that analysis will give real component ratios for evaluation. When the additional phase is solid, the problem of liquid wetting of the solid or retention of liquid on the solid may be very important. This is the classic problem encountered in studying solid/liquid phase diagrams; the method of "wet residues" or the use of tracers from the liquid phase will usually enable a reasonably accurate evaluation of the composition of the solid to be made; the possibility of forming hydrates, solvates or molecular adducts should not be ignored. From the ratio of the components of the solid after correction for liquid phase retention, it is possible to arrive at the true composition of the solid itself and thus to derive a material balance at the multiple phase point.

Once the compositions and equilibrium points of the multi-phase system have been determined, it is possible to evaluate the utility of operating in the multi-phase region.

Aspects of phase separation and recovery of the high concentration phase for stripping and return of reagent are more complex. Thus, in the multi-liquid phase region the nature of the dispersions of the three phases as a function of process and operational parameters needs to be clarified; for example, the dispersion may be of a multiple type which may require primary and secondary coalescence zones, or two types of equipment. Considering "light, intermediate, heavy" phases, the primary and secondary phase separations would thus need to be evaluated, determining which is the continuous and which is the dispersed phase, and the best circumstance for coalescence and separation of phases.

When the additional phase is solid, an aspect of major importance is the wetting characteristics of the solid, i.e. whether it remains in the interface, above or below the demarcation line, or is wholly in the heavy phase or wholly in the light phase. In every case phase separation and recovery must then be considered. In some cases the type of contacting during the additional phase generation may be very important, also the mode of feeding the reacting phase may need to be considered in connection with wetting of the solid phase formed.

Finally, equipment selection for programmed contacting and phase separation needs to be carefully evaluated; particularly the preferred mode of solid separation and recovery may require an in depth analysis, and an attempt at predicting problems so as to anticipate how best to prevent these or to cope with them.

The above exposé, though brief, is sufficient to stress the point made that an inter-disciplinary group study is the only way in which this area can be extended for useful, fruitful application.

II. The LLX Separation—Chemical Feasibility

LLX—Various Viewpoints

Student/teacher, researcher/developer, manager

From the teacher/student viewpoint LLX is a unit operation, parallel to distillation, which has derived its formalized format as a two-phase operation by analogy with distillation.

From the researcher/developer's point of view, however, LLX should not be just a formalized unit operation since it is a separation tool which can be moulded and fitted to the requirements of each case; in other words it is not "given by" the system, it is "imposed on" the system as purpose demands.

The manager's case is an integrated viewpoint; a central question is, which disciplines must be brought together for the successful design and implementation of the LLX process?

Even though these viewpoints seem to differ, they must follow a common line leading from student through to manager, since that is clearly the line of responsibility and personal development. What is it then that can connect these viewpoints? The key must lie in the freedom of imposing two liquid phases on the given system. This is not an arbitrary freedom, however, but it is the "freedom of the informed", and must be based on the realization and appreciation of possible interactions within a system; the selection of one set of such interactions which can lead to at least two liquid phases will then permit and promote separation. This should therefore be the line that the teacher must put forward in the student/teacher relationship, which the researcher/developer will then utilize, and the manager will be aware of, so that continuity is established.

LLX is an ideal area for exemplifying interactive or integrated systems on various levels. First there are the molecular interactions, i.e. those interactions that are the basis for the two phenomena in LLX, namely the achievement of demixing or prevention of full miscibility, and the achievement of a tendency to distribute; then there are bulk interactions, e.g. concentrations, which influence transfer; finally there are process interactions, stream integration, material recycling etc. Optimization of all these aspects is complex; since cases differ so much it is necessary to be flexible in approach, this makes a formalized approach rather difficult to present. Perhaps the problem can be expressed in a different way, as follows. For an established two-phase system, according to its type, it is fairly straightforward to define the basics which need to pass from teacher to student; similarly for the researcher *per se*, the areas to be studied can be selected and defined so that all other areas are legitimately outside the scope of the work; however, there are other types of researchers, perhaps we can call them the problem solvers and the innovators, for them at best the aim may be stated, but the system and solution are open to choice. Since any student may become a researcher of the second type, it is necessary that the difference and the scope come out in the teacher/student interaction, so that it can follow on, also in researcher/developer interaction through to the managerial aspects of process implementation.

Once a two-liquid-phase system has been defined, it can be approached as a LLX system almost without relating to molecular interations; in other words, the skills normally acquired in relation to mass transfer unit operations can be applied without necessitating a definition of mechanisms of transfer.

The case is different when a two-phase system is being sought so as to promote a desired separation by transfer from phase to phase. This requires an understanding of the chemistry in a very fundamental sense, both as regards the mixing and demixing of substances in relation to structure, and the physico-chemical definitions that derive from structure and interaction. The interactions here are between the "solvent" phases even in the absence of the "solute", and between the "solute" and the "solvent" phases. In review one can mention that two-phase systems stretch, e.g. from inorganic melt mixtures, to inorganic melt/organic liquid mixtures, to water/inorganic/organic mixtures to organic/organic mixtures to water/polymer/salt mixtures to water/polymer/polymer mixtures. No doubt one can define other systems which satisfy the two-liquid phase requirement and make them candidates for choice as LLX systems.

In the student/teacher relationship the question then is whether it is better to simplify or to expand by making the student aware of scope, without details, but impressing the necessity of interaction. Interaction here means communication between disciplines and skills—the best that can be sought in the teacher/student interaction is the transfer of an appreciation of scope and also of the disciplines and skills which need to be brought to bear. Quite separately, individuals may acquire specific skills in specific aspects and disciplines, or multiple skills as far as this is possible and expedient. In the researcher/developer interaction the researcher may be specific and limited, but the developer must have the broad view, drawing together the lines of the multiple backgrounds and skills of those with whom there need be interaction. From the developer, therefore, one passes naturally to the manager definition.

This monograph has two central themes. One is to expose and stress the wide variety of two-phase liquid systems that potentially exist and hence the versatility and freedom for choice of such systems for LLX applications, the other is to present the integrated approach to R & D and PD aimed at selecting and using LLX systems for practical purposes. The latter aspect is probably much more readily acceptable by the reader, and indeed occupies the major part of this text. It has been divided into three sections, namely the procedure, the separation and the process.

The former aspect of versatility and freedom of choice is relevant on two levels, one is in the framework of the discipline of LLX itself as a separation technology, the other in the framework of innovation in application.

Since this aspect of versatility and freedom of choice is one which distinguishes LLX from other unit operations for separations, it seems expedient to consider various viewpoints, proceeding from teacher to student to researcher to developer to manager. The latter three are naturally occupied with innovation in application. A fair part of this text relates therefore to how one can analyse the potential of a system for LLX and how one selects the second phase. This generalized aspect naturally overlaps with R & D and PD in each specific case.

LLX—The User's View

The "user" of solvent extraction in fact includes two "users", one can be called the "process designer" who takes advantage of LLX as a separation tool and applies it for a specific purpose, the other is the "process operator", who has to make the separation process work.

For the process designer, LLX is an interesting tool because it is flexible, reversible, low energy consuming, and permits a high degree of separation to be attained. For the process operator, LLX permits operation with liquids in equilibrium systems promising quick response.

The aspects which are important to the process designer are basic ones, like separation and process constraints, process control, solvent selection, solvent stability and losses. Aspects important to the process operator are equipment selection, operating control, and operational constraints.

The LLX flowsheet will consist of two main steps, namely extraction and washing or stripping. In addition, there may be a "back wash" for removing minor co-extracted undesirable components from the solvent phase, or a "back extraction" to remove remaining minor co-extracted components from the product stream. When impurities with high distribution coefficients are present, these may be removed by a "pre-extraction" step, prior to the main extraction, followed by a pre-extract washing step. All these auxiliary transfers and separations are accomplished by recycles and refluxes within the system itself.

In partially miscible systems, solutes can be transferred back and forth, with essentially no net chemical energy effects; the real energy input in the LLX process is due to the necessity to transport the liquids and to contact one phase with the other for mass transfer, which must then be followed by phase separation. The mobility of transfer of solute between phases means that minor changes are rapidly transmitted through the system, so that it is easy to attain steady state, but it is also easy to go out of steady state.

The experience of users of LLX will be personal and specific, each relating to his/her own circumstance, but it is certainly possible to generalize the lessons learned.

When the solvents are selective but not specific, the processes may encompass multiple separations; each separation requiring a limited number of stages, and often showing a sharp concentration profile in at least one of the components being transferred.

Design of LLX Systems

In order to approach the design of liquid–liquid extraction systems in a logical manner, one needs to consider where such systems are being used and how these uses could be extended; also where extraction is not used but could well be expected to find application.

In its most fundamental sense, liquid–liquid extraction implies one thing only, namely mass transfer across a liquid–liquid boundary. Inherent in this basic concept lies the technological definition of liquid–liquid extraction as a means of separation and purification, by combining mass transfer with phase separation.

Classification of LLX systems may be end-oriented, being based upon the type of separation desired, e.g. concentration, recovery, purification, or upon the type of material being transferred such as metal recovery, acid extraction, salt separation. It may also describe the reactions being promoted by the mass transfer, e.g. metathetic reactions, double salt decompositions, equilibrium shifting. A different type of classification would relate to the physico-chemical nature of solute/solvent interactions which form the basis of the mass transfer and hence of the separation achieved.

The description of solvent extraction systems has two distinct facets, the one operational, the other relating to solvent composition and quality. Both are limited by the tendency of the solute components to distribute between the two liquid phases.

Operationally, the format of the system is simple and quite general. Schematic flow diagrams for these general variations, showing also the direction of mass transfer, are presented in Figs. 17(a), (b), (c) and (d).

Regarding solvent composition and quality, the options are much broader and are also much more specifically related to the nature of the mass transfer. In the simplest cases, the solvent will consist of a single component. The next simplest system is composed of an "extractant" and a "diluent"; the diluent is regarded as "inert" and is assumed to influence only the physical characteristics of the system, particularly viscosity, but also to help solubilize the extractant–solute complex. However, since diluents are never wholly inert, active diluents may be selected by intention. Secondary diluents or modifiers may be added as well, thus leading to less simple solvent systems, where interactions within the solvent phase provide a gradation of properties making finer separations possible.

The single solvent case offers small room for manipulation of the system, since it alone must meet all process and operational requirements. In other words, it must satisfy all aspects that will lead to an overall viable system, i.e. specificity, capacity, solubility, mass transfer, phase separation and cost. Systems composed of an extractant and an inert diluent immediately offer more possibilities in selection than does a single component. Thus the extractant itself can be optimized without changing the overall physical properties of the compounded solvent; synthetic organic chemistry has a large part to play in designing extractants for specific purposes. While the inert diluent has been considered mainly for operational reasons, the use of a diluent does immediately provide the system with an additional degree of freedom, by permitting the extractant concentration to be varied at will. The multi-component, interactive solvent system has been mentioned in recent years, but its usage is not wide spread.

Technologically, solvent extraction is a separation tool, hence the measure of its success in a particular application is the degree of separation attained. However, the quality of the separation cannot be the only yardstick; obviously, quantitative aspects of recovery and efficiency, as well as economic considerations of cost, must be satisfied.

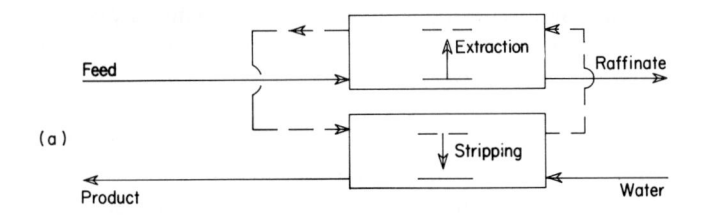

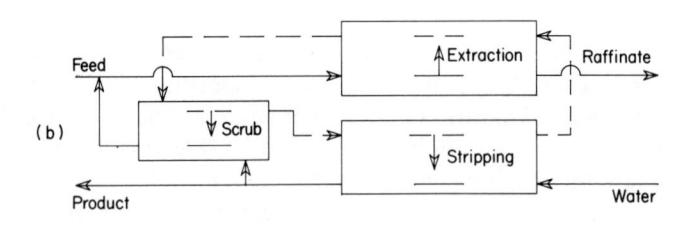

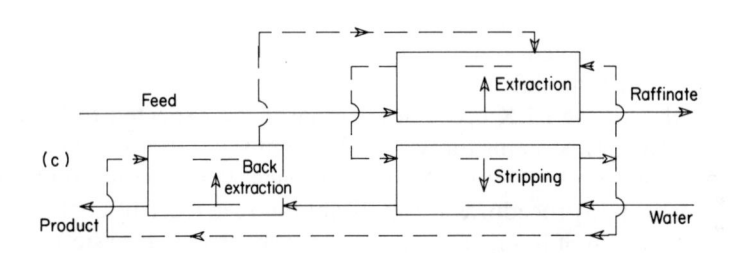

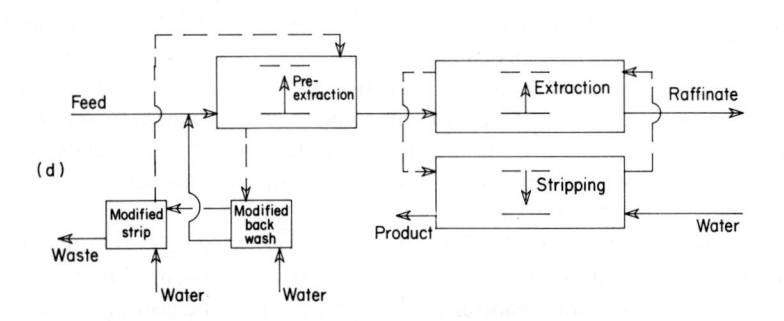

Fig. 17 (a), (b), (c), (d) Back and forth transfers in LLX

System modification or adaptation will usually be motivated by process or economic advantage. Process advantage will derive from better separations, specific separations, or easier separations. Economic advantage will relate to less solvent per unit separated, less costly solvent, reduced losses, better phase separation, faster kinetics, less hazard, lower investment, etc.

For single component systems, the only modification possible is to select a different solvent. For reagent–diluent combinations, the diluent may have to be selected for reasons not directly connected with the reagent/solute interaction. Thus, for example, solubility in the solvent phase may influence the diluent choice, apart from the obvious aspects such as viscosity and interfacial tension.

In designing a liquid–liquid extraction system, an important consideration is the generality of the case. An unfortunate aspect of specific reagents is that they teach little of a general nature. On the other hand, a general approach to solvent systems, where separation is controlled by selection of process parameters or where the reagent can be selected from a graded family, has much to be said for it. In the former case, sensitivity of distribution coefficients to a physical parameter such as temperature, or to a physicochemical parameter like pH, can be utilized for sequential separations. Similarly, a family of extractants can form the starting point for fitting an extractant to a specific purpose, while keeping the concepts general. Another approach to solvent systems of broader applicability is by inter-play of reagent and diluent modifier.

Analogy with existing LLX processes can hardly be expected to be the basis for designing new systems for widely different separations, hence it is necessary to explore other areas. A first approach would be to consider which other separation procedures are applicable and can teach by analogy.

The great attraction of interactive, multi-component solvent systems lies in the possibility of setting up a continuous solution scale, stretching from water on the one hand through to a non-polar, unordered organic system on the other, with all the variations and grades in between. Thermodynamic properties relating to activities, formation constants, stability constants, donor–acceptor systems, heats of solvation, acid–base scales, etc. can be drawn on and pieced together. In this way, one may come to see which property is best exploited for attaining separation, how materials to be separated differ among themselves, what type of environment is likely to lead to separations, and so on.

Liquid–liquid extraction, by definition, requires bringing two liquids into contact to promote mass transfer, and then to separate them. An important consideration then is the residues of each phase left in the conjugate phase.

Similarly, aspects of solvent stability, from economic, environmental and technological points of view, may be determining. The importance of solvent losses from the economic point of view is fairly obvious, but less appreciated, perhaps, is the need to distinguish between chemical process losses and physical, technological losses since these have a different bearing upon the system selection.

Another aspect may depend upon the end use of the separated or purified material, related to the seriousness of the presence of residual solvent components in the product.

Auxiliary operations may sometimes be very significant, particularly where the solvent extraction operation is interposed between other steps; these operations may even make it preferable to choose a solvent system which by itself is not the optimum choice but which can be integrated with advantage into the whole.

There are various ways in which one may come to be considering liquid–liquid extraction for a particular separation. In some cases the current separation procedures may be inconvenient, inefficient, energy intensive, or generally costly. In other cases, one may have considerable technological knowhow in the area and be seeking new applications.

Almost invariably one must start from the separation itself and then come to the solvent system. Indeed, practical results will rarely be attained by developing a reagent, or by synthesizing or compounding a solvent and looking for likely applications. The chemical literature is full of examples of studies of special solvent systems which have been tested with a variety of solutes, but which are never usefully applied. The reason for this may well be that all these cases remain separate without being generalized by identifying and classifying their mode of transfer, the degree of hydration and solvation, or their donor–acceptor relationship, so that each isolated case will fall into an ordered pattern with others and thus become conceptually available for useful consideration.

The design of liquid–liquid extraction systems can be handled only on an inter-disciplinary basis, combining together, for example, the approaches of non-aqueous solution chemistry, of complexation, solvation, acid–base relations, solvent polarity, the phase rule, thermodynamics, synthetic organic chemistry, and so on.

There are only two questions to be asked in approaching a problem: firstly, what is the nature of the material requiring separation, and secondly, which of its characteristics can be exploited to facilitate this separation?

LLX Strategy and Constraints

In process development the strategy will be to approach the aim on as broad a front as is compatible with the external constraints. Options should be kept open for as long as practicable, i.e. provided doing so does not hinder project continuity but continues to bring constructive results. Definitive decisions on variable aspects must be taken as soon as the returns on additional work become marginal, and/or in compliance with the imposed constraints.

The primary constraint imposed on a development project will often be timing; however, investment, scale, integration into an existing facility and overall development plans will all have bearing on the process development strategy employed.

As in all cases of process development, so too when developing a LLX process, it is necessary to arrive at a conceptual model at an early date so that all the operations fit together to attain the aim.

LLX itself entails mass transfer without change of state; by definition, mass transfer is from one liquid phase to another, so at least two liquid phases must be present in the system. This does not imply, however, that additional phases may not be present too. Thus, for example, a third liquid phase may be formed in process; a solid phase or phases may participate in the overall transfer or transformation; usually condensed systems at atmospheric pressure are considered, but this is by no means necessarily so, as the vapor phase too may have a function in the overall scheme.

Basically the LLX process will consist of an assembly of mass transfer operations entailing transfer of solutes back and forth between phases, each transfer achieving a desired separation, so that the sum of all the transfer steps together gives the overall separation and recovery aimed at in the process.

The strategy of process development must be planned in such a way as to arrive at a realistic proposal for attaining the separations and recovery desired, by manipulation of the system variables in such a way that the technological whole is consistent and operable. Furthermore, this must be achieved within the imposed constraints of timing and expenditure. Since this will require compromise on all levels, the overall strategy must aim at attaining some acceptable defined degree of certainty, while locating and acknowledging the areas and degree of uncertainty. This is essential to facilitate decision-taking by management.

In liquid–liquid extraction processes a typical set of significant parameters will usually encompass at least the following:

a. Chemical:
   Aqueous composition and the degree to which this can be varied
   Concentrations of feed components
   Solvent selection, type and composition
   Distributions and separation factors of solutes
   Mutual miscibility limits

b. Operational:
   Flow rates
   Temperature
   Pressure
   Physico-chemical characteristics of phases
c. Process:
   Mass transfer/separation steps
   Solvent quality, stability
   Product(s) concentration, quality
   Effluent(s) composition, concentrations, quality
d. Technological:
   Interactions of (a), (b) and (c) on equipment selection
   Interactions of (a), (b) and (c) on process economics

Additional parameters may relate to a particular case *a priori*, or may come to light as process development proceeds.

In practice, a LLX two-phase system will often consist of more than three components, hence rarely can a simple ternary diagram be expected to represent the phase diagram of real SX systems of practical significance. On the other hand, multi-component systems are difficult to present in simple, usable, graphical forms; nevertheless, it is essential to consider the phase diagrams of such systems if useful separation processes are to be developed. Within the framework of imposed constraints as mentioned above, LLX process development strategy demands, therefore, that limiting experiments be designed which:

(a) can be performed simply and rapidly;
(b) will show up the major interactions of components in the system;
(c) will give a degree of certainty as to continuity of behavior of distributing components, or will show up maxima and minima in separation factors;
(d) will permit rapid collection of data for primary material balance, heat balance, etc.;
(e) will show up problem areas which cannot be solved within the battery limits of liquid–liquid extraction alone, but will require auxiliary steps, using other unit operations, for their solution.

Since it is extremely important to proceed on a broad horizontal front by tackling all aspects of the process in parallel, one should not do an in-depth study into any one aspect until one is sure there are no bottlenecks which may invalidate the process as a whole. The order of work, therefore, is first to select the solvent (reasonably good solvent is sufficient, a better one can be sought later). The second step is flowsheet delineation, giving essential separations and a material balance for major components. At this point, it is essential to prepare an "order of magnitude" cost estimate for an acceptable scale of operation. This cost estimate must be made by close cooperation between the costing and the process development engineers. Battery limits must be defined and the sensitive cost areas shown up, to serve later as a directive for the process development effort.

If the costing is encouraging, the process development will proceed to the next stage, for example:

a. Definition of feed composition, variability limits.
b. Limits for LLX operations, number of stages, aqueous/solvent ratios, minimum-maximum.
c. Yields, maximum vs. product quality.
d. Solvent stability.
e. Mass transfer rates.
f. Mixing and coalescence tests, for all differing pairs of solutions.
g. Solvent recovery or elimination from product, from raffinate.
h. Analytical procedures, process control, physico-chemical data for all streams.
i. Continuous operation, to confirm the process definition.
   Various aspects must be considered here, for example, the minimum test scale vs. the degree of certainty required;
   filling of equipment, start-up, steady operation, stopping, restarting;
   production of materials for further testing, e.g. product for quality definition and market comparison, also effluents for waste disposal evaluation.
j. Materials of constructions.

The process development program can in part be done on laboratory scale, using separatory funnel type tests for simulation of steady state operation, but in part continuous operation, say on bench scale, will be required.

The aim of process development is to arrive as rapidly as possible at acceptable guidelines for process implementation. The more comprehensive these guidelines are, the more well based will the approach to process implementation be.

Guidelines for process implementation require at least the following: process description, flow diagram, stream compositions, material balances, elements for heat balances, analytical procedures, principles of process control, basis for equipment specifications, selection and sizing, materials of construction, basic safety guide, effluent quantities and compositions.

There are a number of other operational aspects of importance, e.g. start-up and shut-down procedures, response to operational changes, ease of true sampling, etc.

Therefore, selection and design of liquid–liquid contacting equipment requires close collaboration and understanding between the process developer, the design engineer and the plant engineer, in order to comply with initial constraints but having due regard to subsequent start-up and operation.

Modes of Studying LLX Systems — Specific Approaches

A simple system can be studied in a simple manner. Thus in a ternary system—water/M/solvent—if one is to compare solvents one can construct a ternary diagram using the titration procedure for obtaining the limits of miscibility, and/or by contacting known mixtures, allowing phase separation, and then analysing the equilibrated, conjugate phases. Instead of constructing a ternary diagram one can construct an equilibrium curve as a function of concentration of M in the two phases, provided mutual miscibility of water/solvent does not change appreciably as a function of M concentration.

As the system becomes less simple, this procedure becomes more cumbersome, e.g. in a quaternary system—water/M,N/solvent—one needs to use a family of ternary diagrams, or a three-dimensional diagram, or a family of equilibrium distribution curves as a function of selected parameters, held at successive levels. If one is checking solvents for comparison in such a two-solute system, the data base required becomes rather extensive.

For more complex systems a different approach is preferable. Such an approach will utilize procedures well established in physical chemistry, physics, and applied mathematics, i.e. the "by inspection" approach. This assumes an appreciation of thermodynamic concepts in multi-component, multi-phase equilibrium systems; in particular it requires an *a priori* evaluation of the significance of the various equilibrium constants which must be satisfied, and which together constitute the distribution coefficients which summarize the system.

One can make a list of all equilibrium constants likely to apply, then mark off those that are clearly of minor importance and so arrive at a limited number of significant ones. These can then be fixed so that parameters which control them are no longer variables. In effect this simplifies the system so that fixed points can be selected for comparison.

This approach has been well established for use with ionic systems. There does not seem to be any reason not to extend the application to molecular, complexed systems, provided the correct limiting equilibrium constants can be selected and fixed.

In a sense this leads to a unified type of approach to LLX systems. The basic unifying approach, of course, is the Gibbs Phase Rule, which previously was applied more to multi-component, multi-phase, but *single*-liquid phase systems.

The Phase Rule provides a very practical framework for looking at multi-phase systems at equilibrium, whether there are multiple liquid or solid phases present or both, and whether one takes the vapor phase into account too, or considers only condensed systems, ignoring the vapor phase and hence losing one degree of freedom.

Distribution coefficients are the accepted manner for expressing the tendency of system components to transfer between two phases. The distribution coefficient is the resultant of all the equilibrium constants applicable to the components in the system, all being related to component activities, which are usually expressed more simply as component concentrations in the respective phases. Similarly, separation factors are the ratios of the distribution coefficients of the components being separated. Since LLX is essentially an equilibrium process, entailing a multi-phase, multi-component system even in the simplest case, the Gibbs Phase Rule must apply:

$$F = C - P + 2$$

where: F = degrees of freedom, C = number of components and P = number of phases.

A number of primary concepts can make these multi-component multi-phase systems quite manageable, so that one can go ahead to select and devise a LLX system to achieve some desired separation without necessarily being able to describe clearly the state of the organic phase. One such basic concept is that at equilibrium the activities of each component in all the phases is the same, hence any phase can be selected to define the activity of a particular component of interest. This means that two different systems at equilibrium are fully comparable with respect to specific components provided they are at equal activity in the two systems. This constitutes a strong tool for selecting systems on the basis of real comparisons.

In general, the problem of comparison is difficult when the system is multi-component and has multiple phases, since the validity of the points being compared is important. The Phase Rule approach offers a lot in such circumstances, since invariant points can be defined and used as a yardstick for comparison. Normally distributions are expressed as function of concentration. This can of course be misleading when there are interactions in a system; consider the effect of a salt with a common ion on an acid which is extracted into a solvating solvent. Clearly the concentration of H^+ is not alone determining nor is the total anion alone determining; it is the ionization constant which determines, since it takes both into account and since the un-ionized compound is extracted.

Driving forces for transfer may have many facets, and in certain cases these may not be clear *a priori*. The driving force for transfer is a thermodynamic force. Let us consider a two-phase system; let us further assume that one phase, say the aqueous phase, is saturated with respect to the component to be extracted; in other words, were it not for the transfer to the solvent phase the compound would precipitate. If the compound is being formed at the same time as it extracts, then the distribution coefficient is controlled by the solubility product, provided the aqueous phase remains saturated, i.e. provided there is some small quantity of the precipitated solid compound present at all times. If now it is desired to compare the behavior of various solvents or solvent compositions, this is a good point at which to determine extractions and separations for comparison. Similarly, for acid–base reactions, half-titration pH is a very useful tool for comparing systems. There are other cases in which, for example, vapor pressure may be a reference for comparing systems, particularly as a measure of equal water activity. In a system where it is not possible *a priori* to identify fixed points as a yardstick for comparison, a different approach may be to use a mixture of the untested system with a tested one, thus showing up interactions which were perhaps not anticipated.

In LLX the problem of driving force for transfer, from aqueous to solvent then back from solvent to aqueous, is a central theme. When the same equilibrium controls in both directions of transfer, there must always be a loss in concentration in going from aqueous through solvent back to aqueous. For recovery of low concentration values, say from an aqueous waste, this may therefore have no sense, since the volume of the final solution will be greater than the starting waste. In such a case, therefore, some mode of changing the driving force must be sought. On the other hand, for removal of high concentration values from an impure environment, a loss in concentration may be offset against the gain in purity, hence operation with the same equilibrium curve for distribution in both directions may be a feasible mode of approach. There may be cases where the separation between solute and solvent by non-LLX approach can be considered. There are various possibilities: evaporation of solvent, precipitation of solute by reaction, precipitation by addition of a second solvent which changes the characteristics of the solvent, membrane separation, adsorption of solute, etc. The viability of such procedures must be evaluated for each case. Where a low concentration solute favors an organic solvent in distribution, it may be fully acceptable to use LLX only in one direction and evaporation of solvent in the other. In such a case the solvent chosen should have a reasonably low boiling point and as low a latent heat of vaporization as possible. This approach may apply, for example, in removing low concentrations of organophilic components from waste waters.

The use of a chemical driving force is a very powerful tool. Let us take the simplest case of metal ion extraction by a carboxylic acid. The controlling equilibrium is:

$$M^{2+} + \overline{2HR} \rightleftharpoons \overline{MR_2} + 2H^+$$

In order for the reaction to proceed to the right, i.e. for M to pass to the solvent phase, H^+ must be removed from the aqueous phase. In other words, pH control will be required, by adding a base, thus losing the H^+ value unless the base of the same cation can be used. In the reverse direction, where acid is added to return the M^{2+} to the aqueous phase, the limit will be the available concentration of the regenerating acid and the solubility limits of the salt thus formed, i.e. the concentration of the counter ion in the aqueous phase to give $M^{2+}X^{2-}$ will be controlling as well as the extent to which the hydrolysis constant is favorable as a function of concentration, that is:

$$M^{2+}X^{2-} + 2H_2O \rightleftharpoons M(OH)_2 + H_2X$$

In practice, in order to attain such high concentrations it may be necessary to recycle the aqueous product in order to preserve a favorable phase ratio for subsequent phase separation.

The difference in driving forces for extraction and stripping can be approached by a reversed procedure, e.g. a common ion effect in extraction where the ionization constant is controlling. In this case if, for example, we have AY and BY where BY is not extracted to any degree, then BY promotes extraction, while stripping is promoted by the absence of BY. The common ion effect is important in a solvating solvent system, e.g. $MgBr_2$ extraction from a $MgCl_2$ system, or $MgCl_2$ extraction from a NaCl or $CaCl_2$ system.

Another mode of controlling the driving force is to change either temperature or pressure so as to affect some factor of the overall equilibrium constant. Temperature can be a powerful tool in acid/base ion-pair formation; pressure is rarely used in LLX but is the basic step in supercritical extraction. Similarly if solubility is affected by temperature, it is possible to move between two temperatures in an invariant system and thus to cause the equilibrium to proceed in the desired direction by virtue of the change in concentration caused by temperature change.

The effect of minor components in changing distributions of major components was studied in the 1950's and was called synergism. Usually the minor components were involved in completing the coordination sphere of the solute, e.g. by replacing water or the solvating bulk solvent itself.

The contribution of "modifiers" in various systems, or of compounds which are said to be "synergistic", provides an entry for study into complex systems. Synergistic cases may also be of significance in transfers that are kinetically controlled in practice, i.e. systems in which the rate of the formation of the complex is determining.

With protic solvents such as alcohols for acid extraction, the alcohol phase will contain a higher acid to water ratio than the equilibrium aqueous phase, presumably due to the fact that the solvent in part displaces the coordinated water of the acid transferred. There are cases of highly hydroxylated acids where this is not so, presumably because solvent cannot take the place of water in a highly reticular hydrogen-bonded system of larger molecules.

The importance of water in LLX systems may be extremely far-reaching. Thus in the metathetic reaction of the type

$$MA + HB \rightleftharpoons MB + HA$$

the presence of water greatly facilitates the transformation. It would appear that the water level in the system should be such that a saturated solution of the MA and MB is attained at whatever H^+ level one desires to operate. This is an invariant system at fixed temperature and acidity, so the level of water in the solvent phase is fixed, due to the water activity in the system, while the aqueous phase acts as transfer medium for the anions.

If one looks upon a multi-phase metathetic system as a set of ternary systems with common sides and/or apices, one sees that one has in fact a continuum. To the extent that solubilities of the components vary from phase to phase, it is clear that the presence of the second liquid phase may move the system to a more desirable region for reaction. This is essentially what happens when an added component "breaks an azeotrope" or "eutectic" and thus permits separation to be achieved. In LLX, too, an added component may move the system to a more favorable area for separation.

How to Approach a LLX Process Study

This approach is based on the assumption that initially only the minimum of relevant data need be obtained in order to arrive at the first outline of a workable scheme. In the next stage details and branchings will be filled in. This is, therefore, not the same approach as is used when a defined system is being studied exhaustively.

Given: feed composition and range.
Solvent system selected: reagent, diluent, modifier.

Extraction: limiting condition for feed vs solvent composition, i.e. vs reagent/diluent/modifier ratios.
Limiting conditions: large volume aqueous, small volume solvent.
Distribution of major component: usually two contacts in either direction should be done, particularly where solvent components distribute differently between two phases. This gives a good idea of equilibrium at the feed end,

i.e. equilibrium between feed and loaded solvent, provided $Aq_{(2-1)}$ and Aq_0 are not too different. This is shown in Fig. 18. It is usually required also to know equilibrium at the lean end, hence it is recommended to continue cross current as shown in Fig. 19, until

$$S_{[1-(n)]} = S_{[1-(n-1)]}$$

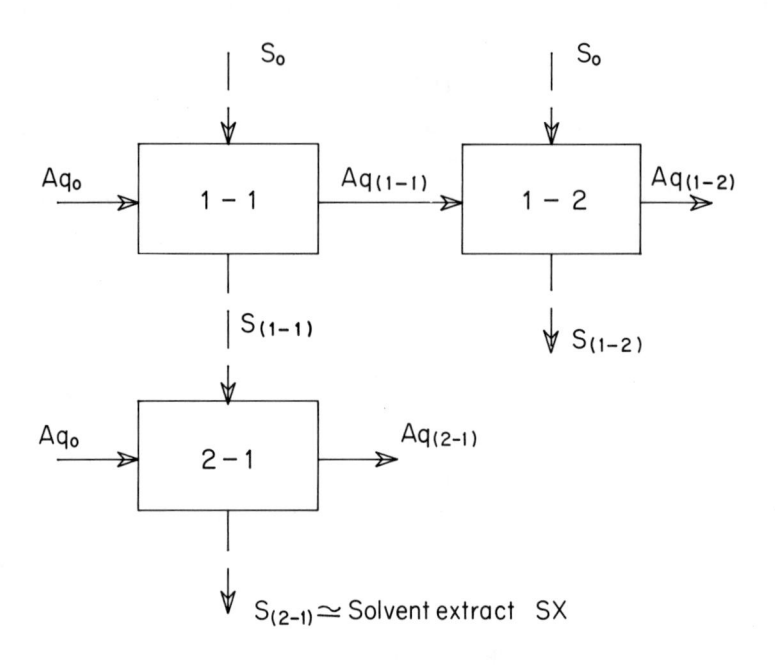

If:

$$Aq_{(2-1)} \longrightarrow Aq_0$$

Then:

$$S_{(2-1)} \longrightarrow SX \quad \text{i.e. Limiting equilibrium}$$

Start numbering stages at entry
of aqueous (heavy) feed phase.
   Row Stage
    (1 – 1)

Fig. 18 Extraction, limiting conditions, equilibrium at feed end

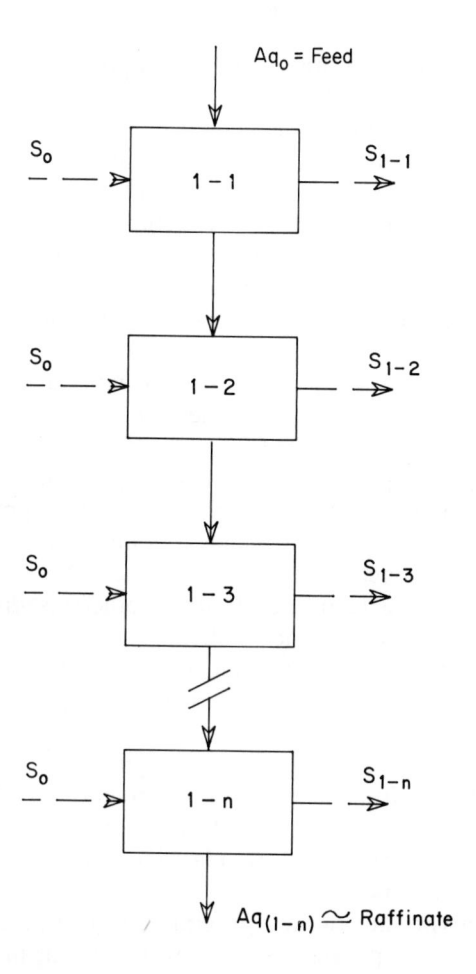

When

$$S_{(1-n)} \simeq S_{(1-(n-1))}$$

Then

$$Aq_{(1-n)} \longrightarrow \text{Raffinate}$$

Start numbering stages at entry of aqueous (heavy) feed phase

Row Stage
(1 – 1)

Fig. 19 Extraction, cross current, equilibrium at raffinate (lean) end

Thus we arrive at another equilibrium point $Aq_{(1-n)}$ (depleted aqueous feed) vs $S_{(1-n)}$ (approaching feed solvent). A comparison of the two distribution coefficients, i.e. at the loaded and at the lean end of extraction, will indicate whether they show concentration dependence—if so, it will be necessary to determine intermediate points.

The solvent composition too can be tested in a similar manner, by selecting levels of reagent and modifier. Probably two points are sufficient to show up relevant parameters; however, three or four points are required in order to define trends as a function of these parameters.

Stripping: Loaded solvent is now stripped with water; again distributions are determined for the loaded and the lean ends, as shown in Fig. 20. The conjugate phase $W_{(2-1)}$ and $SX_{(2-1)}$ will give a good indication of the loaded end of stripping provided $SX_{(2-1)}$ and $SX_{(1)}$ are not too different. Cross current strip with water for n stages will approach full strip; when $SX_{(1-n)}$ and $SX_{[1-(n-1)]}$ are almost equivalent, the end of stripping has been reached, hence the conjugate phases $SX_{(1-n)}$ and $W_{(1-n)}$ give the distribution at the lean end. If the coefficients differ at the two ends, additional points must be included.

In the system selected, the same check should be made at least with the solvent composition at the two levels of each component tested in extraction, since it may well be that while the effect of composition on distribution in extraction is not great, it may be more significant in stripping, or vice versa.

When the requirement is not only extraction of M, but separation of M from N, then at each step the distribution coefficient of N too must be determined so that the separation factor can be calculated and related to the variable parameters.

During this preliminary test program, the distribution of the basic components of the two phases must be examined, again at both ends of the extraction and stripping. This means that water must be determined in conjugate solvent phases, and individual solvent components analysed in the conjugate aqueous phases. If there is a distribution of water or solvent components, the significance of the experimental procedure in relation to the distribution must be examined, particularly since cross current procedures may be accentuating water transfer, or depleting solvent of a component as a result of higher distribution.

The procedures outlined thus far will give a reasonable appreciation of the limits of the LLX system being considered for the main purpose, and will permit selection of the ranges for a first material balance calculation. Depending on the distribution coefficients and separation factors, the question of transfer stages can be estimated or calculated. This is probably easy with the aid of a computer program, but also not difficult using graphical procedures and acceptable simplifying assumptions.

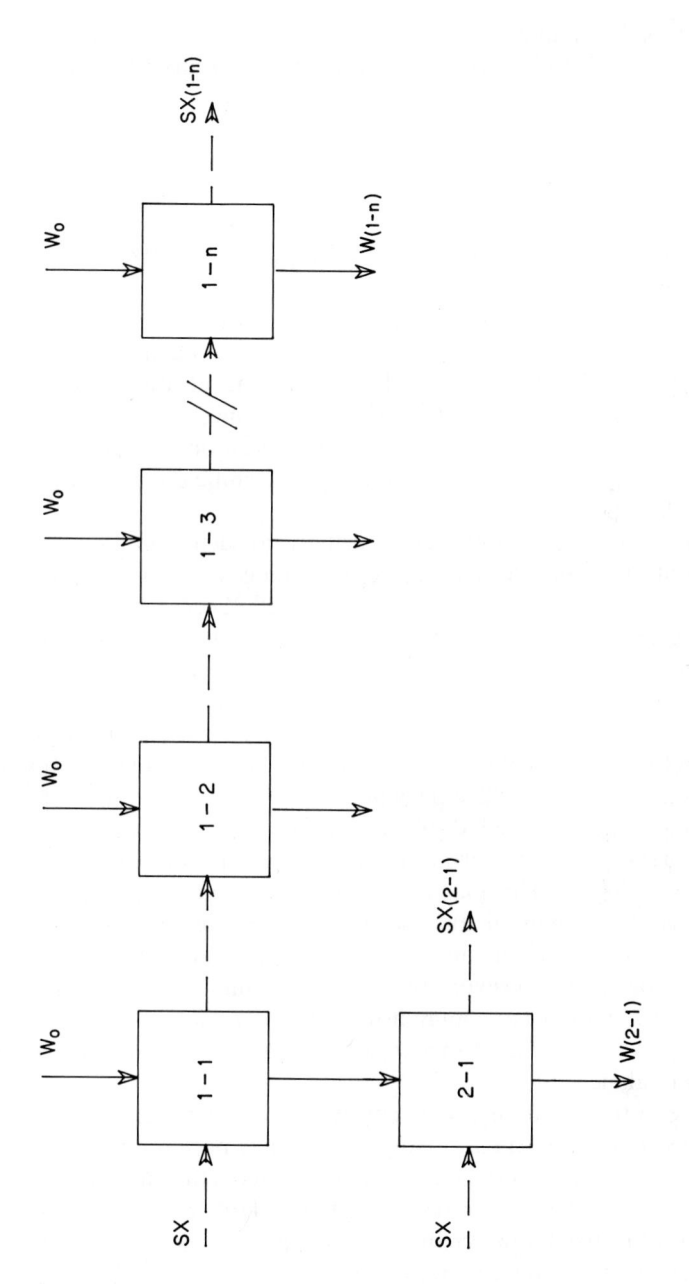

Fig. 20 Stripping

The solvent may in certain cases be selected as a solvating agent, that is to say it should itself also satisfy all the other requirements of the extraction, and therefore not be a compounded phase; this means there is no reagent, modifier, or diluent, since the single solvent takes over the functions of all. In a sense the test program is simplified in this case; it should be clear, however, that in a cyclic operation the solvent will probably never return to the original condition of the dry, virgin solvent. Tests must therefore preferably be repeated with recycle solvent out of stripping, back to extraction.

When the extraction is based on ion-pair formation or on ion exchange, or when a specific complexing agent is used, the extent of extraction will be stoichiometrically controlled, hence the concentration of the reagent in the solvent phase will control the material balance and the phase ratio. However, solubility and physical property limitations, especially precipitations or viscosity, may be the factors which determine practical reagent levels in a specific case.

In a solvating system there is usually no such limitation, but the limit of partial miscibility may be the determining factor, for if the plait point is reached, only one phase will exist hence LLX ceases to operate as such.

Again, when components M and N are to be separated with a solvating solvent, it is very important to determine separation factors at both ends of extraction, and at both ends of stripping. The former will indicate whether there is a transfer dependence of the one component on the other, and therefore whether there is any likelihood of accumulation in a multi-stage system. The latter will show whether additional separation can be attained by a back wash of loaded solvent prior to the stripping.

In a system using a solvating solvent, the importance of water balance may be far-reaching. This may be less so in a system using a diluent which itself has very low mutual miscibility characteristics with water. Indeed, the addition of such a solvent also to a solvating system may help to control the water balance; however, two opposing influences will then have to be matched, namely the commensurate decrease in the distribution coefficient of say M, due to the change in the character of the solvent, versus a better water balance.

When there is a large concentration difference between M and N in the feed stream, and in cases where N extracts preferentially by virtue of the presence of M, or has a higher distribution coefficient than M, it is possible to use pre-extraction for removing the bulk of N before extracting the bulk of M. In principle, the viability of this approach will depend on how well M and N can be separated afterwards in conjunction with the pre-extraction, so as to prevent excessive losses of M while still rejecting N.

Overall, in a LLX process study, it is necessary to arrive at a flowscheme proposal which permits recycling or rejecting streams generated in the system, according to the equilibria and the economics of the system. There is considerable analogy between this multi-operational LLX system and a distillation system comprising rectification and stripping. In a sense the LLX system is simpler, since it is not usually necessary to take the heat balance into account, except in systems which entail acid–base reactions, or possibly where transition points are to be spanned.

Separations can be effected by a multi-operational procedure centering on the extraction, or by a multi-operational procedure centering on the stripping. The determining factors will usually be the relative distribution coefficients and the relative concentrations of the components being separated.

Thus minor components showing higher distribution coefficients in extraction than the main components are conveniently removed in a pre-extraction cycle, while minor components which show a lower distribution coefficient will preferentially be removed in a back wash or scrub step prior to stripping.

Based on this relatively circumscribed process study, a whole flowsheet can be calculated showing both a quantity balance and the anticipated concentrations in the various streams.

Laboratory Simulation — Process Study of a Defined System

First example: acid extraction by a tertiary amine essentially insoluble in water

The tertiary amine will be a commercially available C_8–C_{10} or a C_{12} amine. This amine can be dissolved in various solvents—aliphatic or cyclic saturated hydrocarbons, aromatic hydrocarbons, C_4 or higher alcohols, or combinations of these. The aliphatic hydrocarbon diluent will be the most hydrophobic of the three, followed by the aromatics and then the alcohols which constitute the most hydrophilic, also the most basic systems. Mixtures of diluents will give various intermediate cases, as a function of composition. When a minor quantity of an alcohol is added to a hydrocarbon–amine system, the alcohol is usually regarded as a modifier, since it prevents third-phase formation as a function of the amine/acid ion-pair solubility in the hydrocarbon, and also increases the basicity of the amine.

Once a composition has been selected by experience or from publications, the limiting procedures exposed previously are followed in order to arrive at the distribution diagram for extraction. The present case will be stoichiometrically limited by the amine quantity, hence the equilibrium diagram will be as shown in Fig. 21, as a function of amine concentration in the solvent phase. An operating line can readily be added representing either concentrations "in and out" or phase ratios, since there will be little volume change in these cases. In the graph shown in Fig. 21, two theoretical transfer stages are required.

Let us assume that stripping will be by a competing base, say NaOH, thus:

Extraction: $\overline{Am} + HX \rightleftharpoons \overline{Am.HX}$

Stripping: $\overline{Am.HX} + NaOH \rightarrow NaX + \overline{Am} + H_2O$.

If the loaded solvent is titrated with NaOH, versus pH in the conjugate aqueous phase, this represents the stripping equilibrium curve; if this exact stoichiometry is followed, provided contact time for transfer suffices, the theoretical endpoint will be reached at the stoichiometric pH. Concentration of NaX in the aqueous conjugate phase will depend upon the initial NaOH to water ratio, plus the mole of water liberated by the neutralization.

In this "ion-pair-extraction/neutralization-stripping" cycle, heat will be absorbed and generated, hence in this type of operation a heat balance must be observed. The temperature of the aqueous feed to extraction, and of the aqueous NaOH to stripping may be used for temperature control, or the solvent stream may be heated or cooled as necessary, for steady state operation.

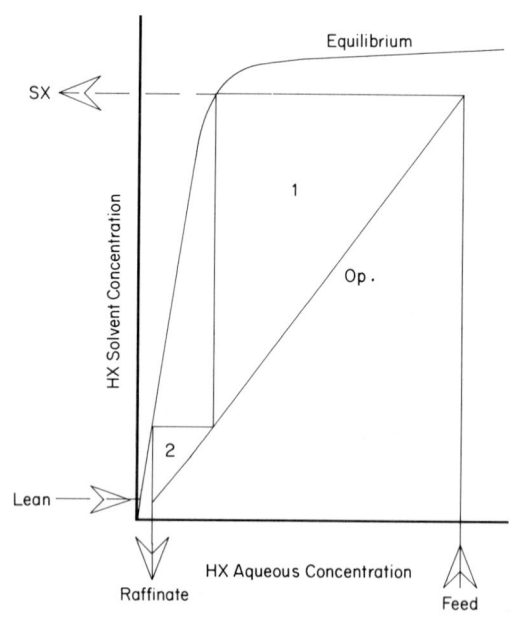

Fig. 21 Amine/acid extraction

Second example: acid extraction by a solvating solvent of limited mutual miscibility with water

Here let us assume the solvent is a C_4 or C_5 aliphatic, primary alcohol. The ternary mutual miscibility diagram will have the typical appearance shown in Fig. 22. The equilibrium curve is presented in Fig. 23.

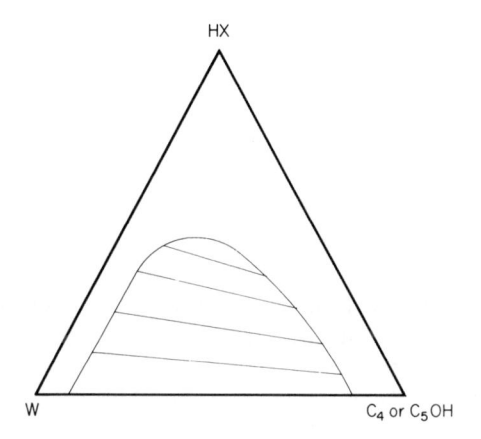

Fig. 22 Acid/solvating solvent—ternary diagram

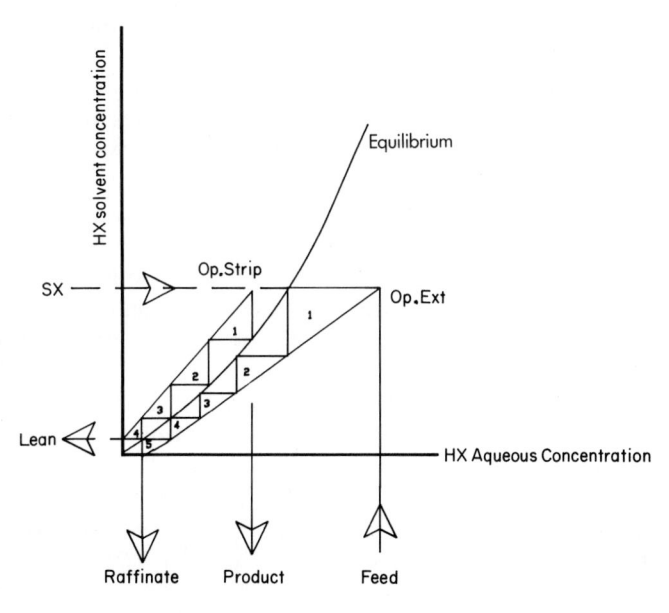

Fig. 23 Acid/solvating solvent—extraction/stripping

Here too, an operating line can be placed on the diagram as a function of "in and out" compositions, both for extraction and stripping. Five stages are required for extraction here, and four stages for stripping. Since the same equilibrium serves both operations, it is clear that for any real number of stages, the strip product will be more dilute than the aqueous feed. Also, because of limited mutual miscibility of aqueous and alcohol phases, the aqueous raffinate exiting from stripping will always contain a limited quantity of dissolved alcohol and the stripped solvent phase will contain some quantity of dissolved water. In general, the water balance here is much more important than in Example 1, which deals with essentially immiscible phases. In Example 2, analyses for all samples must include water content.

In general for both extraction and for stripping, a simulated countercurrent procedure can be followed in a multiple batch mode, using separatory funnels, preferably in conjunction with a centrifuge to ensure phase separation.

In Example 2, mass transfer will be rapid in both steps, while in Example 1 sufficient contact and time must be ensured for equilibrium to be attained, particularly in extraction.

This procedure will confirm the material balance at steady state, i.e. when successive exiting streams of each of the phases are equal both as regards composition and quantity, as well as confirming the calculated number of theoretical stages required for each separation.

This simple approach to defining and confirming requirements can be adapted to more complex systems, such as to the case of an acid/salt mixture with common anion, or to two acids requiring separation.

As additional operations are introduced, it becomes more difficult to use multi-stage batch procedures for confirming a flowsheet and material balance for steady state operation, since feed to one operation must be obtained as product in another and the whole procedure becomes tedious and time consuming. It is relatively simple, however, to set up a modular multi-stage bench unit which can accept feed continuously and thus generate product continuously. The problem here will usually be the control of dispersion type so as to attain acceptable phase separation and coalescence rates, in order to obtain reasonable stage efficiency. There is of course no reason not to use additional modular stages since this continuous bench operation is not the way in which one would determine the number of stages required for large scale operation. The required number of theoretical stages needs to be known, but after that practical stages will largely be a function of equipment selection and the kinetics of mass transfer and phase separation.

An Exercise in Developing an Interdisciplinary Program of Study Aimed at Utilizing Third-Phase Formation

This is a mode of improving concentrations and volume ratios (or volume handling) in LLX systems. The third phase will be a *reagent–solute-rich phase* in the form of a third liquid phase or a solid phase.

An exercise of this type is not directed towards solving a specific problem, but is expected to serve as a pattern for later specified studies while also representing a practical mode of acquiring generalized basic information within the defined scope of the program.

1. Background material:
   1.1 Define system type.
   1.2 Chemical phase data.
   1.3 Approach to flowscheme, using generalized scheme.
2. *Representative case for a specified system type*:
   2.1 Select type of system to be studied.
   2.2 Obtain basic limiting data for system.
   2.3 Fit the data to a preliminary flowscheme material balance.
3. *Technological approach*:
   3.1 Specify aspects requiring study for the type of system flowscheme postulated.

1. Background material

1.1 System types. It is difficult to say how best to approach this stage of the program, since by definition it is the least defined and the most vague.

One approach is to look at possible solute/reagent interactions which in fact will be the same ones that are used in defining mass transfer in LLX systems. Accordingly we have transfer by solvation, transfer by chelation or complexation, transfer by ion-pair formation, transfer by ion exchange.

1.2 The next step will be to examine equilibrium data in order to see whether the appearance of a third liquid phase can be postulated from published data for various types of systems, or to approach the subject in the reverse by noting practical systems where a diluent/modifier combination is used to prevent third phase formation. Thus ion-pair systems are obvious candidates.

Solvating systems too have been described which have this property.

1.3 Regarding flowsheets, as far as the author is aware, only one type process has been described in which the third liquid phase is actually used for flowsheet optimization—this is a two-temperature acid–ether solvating system. Some data have been published on the effect of a modifier on the three-phase zone, e.g. in acid–ether–alcohol systems. These flowschemes and the published criteria and operational options can serve as a starting point for a study of the type envisaged.

2. Representative case(s) for specified system type(s)

2.1 Ion-pair systems, whether the extractant be cationic or anionic, are known to present the "problem" of third-phase formation. These are therefore good choices for the basic study here. It should not be forgotten, however, that a modifier, which prevents "de-mixing" also has a definite effect on the acid/base relationships of the ion-pair systems, hence elimination of the modifier to permit third-phase formation must be viewed, also, from the point of acid or base strength of the reagent.

Solvating systems which give three liquid phases may show considerable variability according to the solvating solvent used. The range covered by the presence of the three phases can be a useful variable, and should be delineated.

2.2 Limiting data for the ion-pair systems selected should cover ranges of concentrations of extractant and solute as well as ranges of that parameter such as temperature, to which the third phase is sensitive. Concepts of the Gibbs Phase Rule should be applied to check for invariant points according to the possibility of fixing parameters such as temperature or concentrations.

Three-phase solvating systems will have a variable number of degrees of freedom according to whether the solvent is uni-component or multi-component, hence limiting data will be considerably more extensive than for invariant or pseudo-invariant systems. The definition of limits (e.g. by ratios or concentration ceilings) may cut down experimentation very considerably.

2.3 Fitting limiting data to a flowscheme and preliminary material balance will probably entail studying a number of different cases. Thus when the multi-phase system can be looked upon as a reacting system, i.e. where some conversion is taking place, it may be expedient to consider such conversions at invariant points. Alternatively it is possible to consider aspects of depletion and recharge, in other words moving back and forth between two fixed points. This can be regarded as an imposed oscillation; there are multi-phase conversion systems that are self-oscillating, such as those which move back and forth naturally between two fixed points.

It is clear that the flowscheme and therefore the material balance will be influenced by the purpose of the LLX step. In other words, conversions or separations or recovery will require different approaches. Operation at a pseudo-invariant point may not be invariant with regard to diluent/reagent;

often this is not taken into account since reagent may be considered on diluent-free basis. Similarly, solvating solvents may be expressed on water-free basis to comply with the pseudo-invariant concept, although on composition basis there is continuous change.

3. Technological approach

3.1 Identifying aspects requiring study for operating a process based on the flowscheme and material balance envisaged demands a high degree of engineering awareness. In any LLX process the basic operating aspects will relate to contacting and phase separation, steady state, operating control, and process control. Even when operation has been postulated for an invariant system, the handling of the extra phase may require an entirely novel approach to contacting and phase separation. When extraction and stripping are run under different conditions it may be necessary to consider whether the third phase is present in a minute quantity simply to ensure invariance, or whether the third-phase solvent composition is changing as a function of solute concentration. In the latter case it may turn out that the phase ratios of L/M and M/H are changing all the time along the flowscheme. This complicates the flow control and hence the operating and possibly process control. The problem may be to find a simple measurement which will readily mirror the state of operation. Here engineering may have to depend heavily on inter-disciplinary involvement; it may turn out that on account of the technological problems the postulated flowscheme is not a working proposition and return to item 2.3 above is necessary to fit in with item 3.1 in a more realistic way. This in turn may refer back to items 2.1 and 2.2 in order to redefine the system. Finally, a return to item 1 and its sub-divisions may be warranted, in a sense starting again—so for a new system this has in effect become a case of trial and error reiterations. To determine to what extent a model can be used to cut down unnecessary cycles, without burdening items 1.2, 2.2 etc., is part of this overall basic study.

Interactive Solvent Systems

The possibility of using a common principle and adapting it to different systems by change of parameter provides a wide basis for development of extraction methods. Various reagent groups fit this requirement. Thus amine extractants give a wide possibility for application because, when the amine ion-pair is used, extraction is possible from highly acidic solutions or from slightly acidic solutions at high salt concentrations.

Also, stripping is possible by different methods, e.g. by changing the acidity (pH control) or by salt formation with a base, or by solvent additive. Thus modifiers can be used to change the effective acidity of the acidic extractants and this facilitates stripping.

Interactive systems are clearly seen in the acid extraction by ethers which form identifiable complexes or solvates (etherates), and where admixture of alcohols act as modifiers, but maintain the characteristics of the ether system.

While amines are primarily regarded as bases, and hence their main application is in extraction of acids, yet amine salts (amine–anion pairs) themselves have been used as extractants. Amine "salts" are good extractants of cations in acidic media as well as in pH controlled media. The cations appear to order themselves essentially according to their basicity or the hydrolysis constants of their salts, e.g. Mn can be extracted when the hydrolysis lies to the right as a function of pH.

$$MnX^{2-} + 2H_2O \rightleftharpoons Mn(OH)_2 + H_2^+ X^{2-}$$

Interaction of minor components with major components can be seen in amine systems where, for example, a minor quantity of a secondary amine will change the shape of the equilibrium distribution curve of an acid in a tertiary amine/acid/water system, at the low acid end (not at the high acid loading end) which has significance for stripping, for example, by temperature rise. Also addition of a small amount of a stronger acid of low molecular weight in an amine/acid system will change the distribution of the major weaker acid, probably because it changes the practical base strength of the non-aqueous phase, as expressed by conjugate pH. This effect exists also with solvating solvents, but here the minor components may also affect the quantity of water dissolved in the solvent phase which is strongly interactive with the distributions. The overall concept must be that activities have changed due to the additions and/or that the extractable species has changed.

In a three-component condensed system with two liquid phases, it is simple to determine the distribution of a single solute and also the mutual miscibility of the two liquids, as a function of solute concentration and temperature. As the number of solutes increases, the situation becomes more complicated since the presence of one solute may influence the distribution of another. This may also be the case in regard to the mutual miscibility of the two liquids, since solutes may cause one liquid to dissolve in the other by virtue of hydrates or solvates, as the case may be; this can perhaps be called "conditional solubility".

In water-containing, partially miscible, two-component systems, the solubility of water in various solvents will differ considerably—still the activity of water in all these solvents will be essentially the same, since the conjugate aqueous phase is water, provided the solubility of the second component in the aqueous phase is low. As the latter increases, the partial vapor pressure of the water changes, so one may no longer be comparing equal conditions. Since partial vapor pressure of water or of the solvent, if it is volatile, is an absolute measure of activity, one can define comparable conditions for various solvent/water systems.

Now let us assume that we have also a solute in solution in the aqueous phase which does not distribute but does affect the water partial vapor pressure of the aqueous phase, then one can set up a comparative scale.

In a countercurrent multi-component extraction system where a solvent extracts real quantities of water while it is extracting solute, and where the solvent has also a real solvent power for water, the water balance can have far-reaching significance in a steady state process, since the distribution coefficient of the water is likely to differ from that of the solute, thus causing concentration or dilution as the case may be.

A system with water transfer in the correct direction can be utilized for concentrating an aqueous phase.

The effect of temperature on water distribution may be very considerable in some cases. The concept of solute/water extraction on "solvent-free basis" gives a scale for comparison of solvents.

Various expedients of transfer can be utilized so as to help balance the water.

Technological Aspects of Transfer from Liquid to Liquid

Now we shall look at the technological aspects as these relate to the desired LLX separation, whatever this may be. In other words, the two-liquid system has been selected, and we need now to determine how the required separation will be attained by the mode of transfer anticipated in the two-liquid system selected.

We need to know whether the system lends itself to operation at an invariant point, so that multi-stage treatment need not be considered, or whether some multi-stage extraction procedure will be required. We need to have an indication of the shape of the equilibrium isotherm, so we can evaluate by inspection the determining factors for devising the flowscheme.

While it is clear that as long as we regard the distribution as attaining equilibrium under the conditions imposed, we need not necessarily know all the equilibrium constants that are being satisfied in this particular case, because the distribution will in fact be the resultant of all the known or unknown constants—we can, therefore, consider the distribution as representing a "practical" equilibrium constant. Any change in parameter may affect one or more of the unlisted equilibria obtaining in the system, and these effects will be reflected in the practical equilibrium constant. This is the reason that distributions under fixed conditions are the simplest way to present the behavior in a LLX system. It should be clear, of course, that this gives a distorted picture, and one may miss completely the importance of certain parameters if this is the way we proceed.

In principle we are interested in the equilibria which interact to control the eventual concentration of components, which then limit the distribution coefficients which express the efficacy of a LLX system. Equilibria which would limit the concentration are, for example, ionization constants, dissociation constants, hydrolysis constants, solubility products, complexation constants, formation constants, stability constants and so on, according to the nature of the components in the system.

Nevertheless the parameters which can be changed in a particular system will be limited by the system itself or by the technology or economics, hence even without breaking down the distribution constant into all the successive constants that satisfy the system, it is worthwhile to determine the trend of the distribution coefficient as a function of realistic parameter changes.

Once the distribution coefficients have been determined it is possible to look at mass transfer (not rates at this point, but quantity) and to determine the number of transfer stages required in going from point M to point N. It is then possible to see how best to arrange the flows in order to serve the purpose. If a computer program simulating the system can be set up it will be extremely easy to check out the variations. These then can be evaluated from the point of view of volumes and stages.

The concept of equilibrium transfer stages developed for distillation and absorption has been applied also to liquid–liquid extraction. A stage-to-stage calculation is the procedure used for multi-component volatile systems when total molar quantities are changing drastically in the two phases. When this is not so, i.e. where passing streams are essentially constant in volume or weight (whichever is the concentration reference used), other procedures, especially the McCabe-Thiele graphical procedure, can be applied. In LLX this is the case when operating at low concentrations, hence the McCabe-Thiele procedure can be adapted to such systems. Problems arise when considerable volume changes take place, e.g. when there is a transfer of water along with the solute, or additional to that transferred with the solute; problems also arise where extraction is concentration-controlled but stoichiometry-limited.

For an unknown system two approaches can be followed for arriving at an estimate of the number of equilibrium transfer stages required over a specific concentration range; one is experimental, by experience and inspection, the other is graphical by making simplifying assumptions so as to obtain a straight operating line and a valid equilibrium line. It is clear that with the introduction of the computer and the possibility of rapid reiterations, the usefulness of simplifying assumptions declines. The experimental approach can easily be applied when multi-stage laboratory simulation is used, as each additional separatory funnel means an added stage; with a modular continuous bench scale mixer settler system the study is probably just as easy. With columns it implies adding a module of height which may be less simple; with centrifugal contactors the number of transfer stages can be calculated only if a correlation already exists, so this is not a primary tool. With an unknown LLX system the simplest approach seems to be the graphical McCabe-Thiele adaptation, using limiting condition equilibria and cross current extractions, where these do not mislead. The essentials for adapting the McCabe-Thiele graphical procedure to multi-component systems is given in Table 15.

Stages can be counterbalanced by dilution in some cases. The importance of stages depends on the rigidity of the separation limits aimed at; their number will be influenced largely by the shape of the equilibrium curve for distribution of components between the phases. This also determines the usefulness of added stages, in other words, the presence or absence of a

TABLE 15. Simplified graphical computations.

Adaptation of McCabe-Thiele graphical procedure for transfer stages.
Requirement:
Straight operating line, i.e. a constant linear relationship between phases flowing through the whole system.
Procedure:
Select two countercurrent streams of components which bear a constant weight relationship to each other. These become the reference phases for expressing concentrations; thus the material balance is calculated on a fixed basis of constant ratio of these reference phases.
Examples:

No. components	4	4
Designation	ABCD	ABCD
Transfer	C from ABC to D	C from ABC to D
Solubilities:		
Mutually immiscible	A,D	B,D
	B,D	A,D
Partially miscible	C,A	A,B
Completely miscible	C,D	C,A
		C,D
Basic streams	(1) AB	(1) B
	(2) D	(2) D or
		D plus defined level of A.

pinch point has basic significance in this evaluation. When distribution is limited by stoichiometry, the addition of stages may soon reach the point of "diminishing returns".

Typical graphs to show dilution vs. stages—amine systems, acid/alcohol systems

Let us take the case of acid extraction, comparing the solvating system in Fig. 24(a) with the ion-pair forming system in Fig. 24(b), so as to see the effect of additional stages in the two cases; for simplicity at this point we shall assume that an alkaline wash is used in both cases for stripping acid from the solvent extract so that the solvent returns to extraction with essentially zero acid content.

Using a simple graphical analysis it is clear that for specified E and F concentrations, the shape of the equilibrium distribution curve in each case determines the number of contact stages required. It is also clear that in an ion-pair system increasing the stages to two instead of only one will double the concentration of solute acid in E, and if desired will also reduce the concentration leaving in R, the raffinate.

In the solvating system, too, we see the same relationship, here in the reverse, i.e. halving the concentration of solute acid in E, will halve the number of stages required for the same raffinate or even a slightly lower concentration of residual acid in raffinate. The real controlling factor is the ratio of feeds, i.e. the ratio of F to S. We have here, therefore, a number of interactive factors, namely distributions, feed ratios and the ability of the equipment to satisfy the stage requirements.

The evaluation will therefore be an economic one as well as a process consideration. Cost of contacting/separation stages and volumes must be evaluated against the implications of higher concentration in the extract and lower residual concentration in the raffinate. Here, intentionally, neutralization was specified for acid stripping from extract, since this essentially eliminates a major consideration in regard to extract concentration, namely its effect on the subsequent acid separation from solvent.

In a solvating system the concentration of solute in the extract is a primary controlling point for the stripping step, as can be seen from Fig. 25(a), since the equilibrium distribution curve is common to both steps, extraction and stripping, hence product will be more dilute than feed in practice.

In ion-pair extraction of acid which is represented, for example, by the following equation for a tertiary amine extractant:

$$\overline{R_3N} + H^+X^- \rightarrow \overline{R_3NH^+X^-}$$

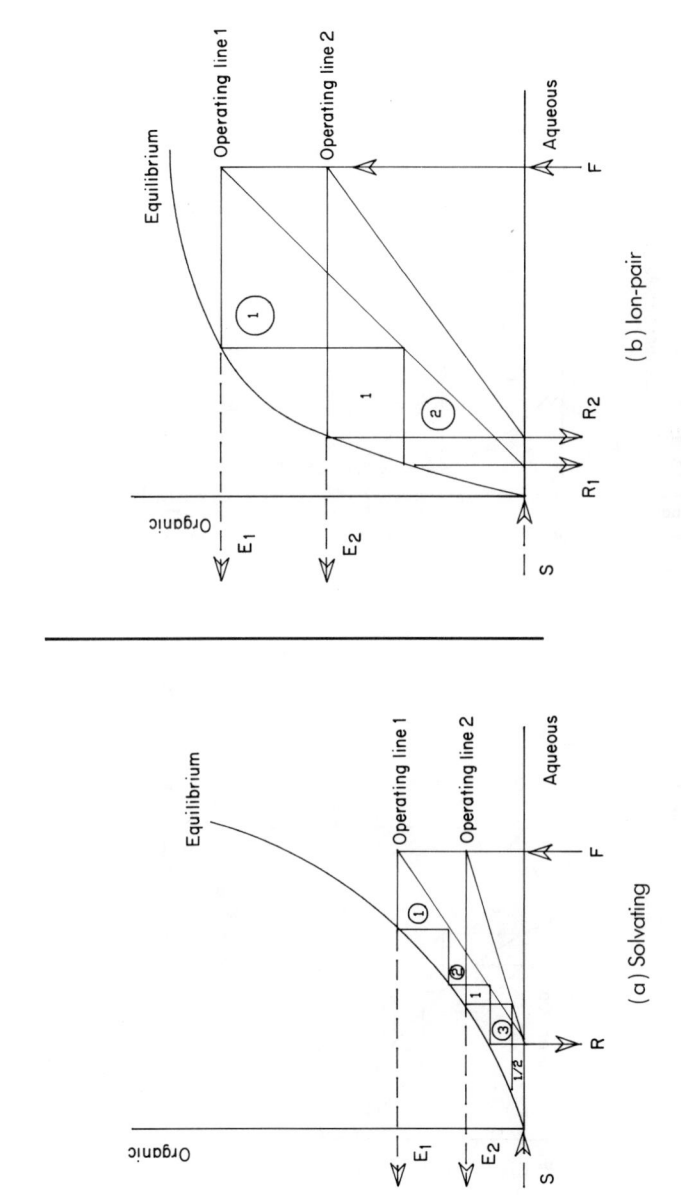

Fig. 24 Acid extraction: dilution versus stages (a) Solvating; (b) Ion-pair

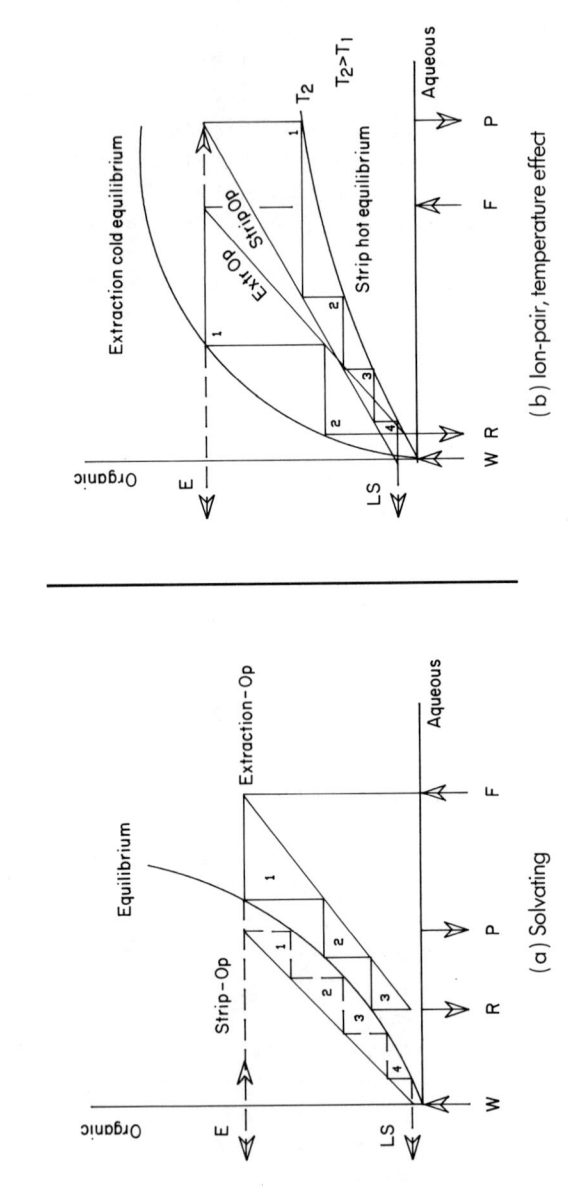

Fig. 25 Acid extraction and stripping (a) Solvating; (b) Ion-pair, temperature effect

reversal of this reaction requires either a competing base or sensitivity to some parameter, e.g. temperature. In the latter case, the equation becomes:

$$\overline{R_3N} + H^+X^- \underset{T_2}{\overset{T_1}{\rightleftharpoons}} \overline{R_3NH^+X^-}$$

where

$$T_2 > T_1$$

The distribution curve therefore changes its shape as the temperature is changed, as shown in Fig. 25(b).

Hence selection of a two-temperature operation permits a system whereby the product concentration can readily equal or exceed the feed concentration, provided sufficient stages are available for the resultant phase ratio.

These and similar arguments can be extended to other cases, e.g. anion exchange, cation exchange, complexation, etc.—in all cases the crux of the matter is the definition of the equilibrium constant which defines the distribution between the two phases, and the identification of the controlling concentration, or the sensitivity of the equilibrium constant to some variable parameter. This is the crux because it is then the basis for selection of the operating cycle with the second phase going through E to S and back to E.

For practical systems, one needs to develop procedures for obtaining data relevant to the case and the purpose, while restricting the scope of data collection as much as possible. (It is true that new rapid instrumental procedures for analysis, and new approaches to data processing, may make this less critical than it was in more tedious days.) The concept of limiting conditions is an established concept in physical chemistry, and can be used very effectively for obtaining equilibrium distributions at predetermined compositions. In conjunction with limiting conditions, invariant systems, or at least elimination of degrees of freedom can be very useful, particularly when comparable situations are being sought. This leads to the use of "reduced" values, for comparisons to be valid, or for data to be comparable.

Another very useful approach in multi-component systems is cross current equilibration systems, particularly when one has major or minor distributing components. From the equilibrium data, one can very easily see whether the high- or the low-concentration end of a multi-stage distributing system is likely to control, and what concentrations one can expect to attain.

Finally one comes to multi-stage, countercurrent simulations. It is clear that equilibrium under limiting conditions, representing essentially true equilibrium with the initial feed, cannot be attained in a practical countercurrent extraction for two reasons: firstly it represents the end of an infinite number of contact stages, approaching equilibrium asymptotically, and secondly because differences in distribution coefficients of various components mean that when finite volumes are used, the limiting equilibrium compositions of all components, simultaneously in extract, cannot be attained—those with higher distribution coefficients than the reference solute will be diluted in the extract, while those with lower distribution coefficients will remain in part in the effluent or raffinate and will be only partially extracted.

The use of analogies in flowsheet delineation is very important and very helpful. The types of steps in a LLX process are limited and repetitive in concept though not in purpose. This makes it reasonably easy to take developed processes as analogy for a new separation.

In any LLX program, starting with a known feed and a known type of solvent, it is possible to block out the steps to be followed, based on the limiting aims and flexibility of the solvent system and the operating parameters.

The scope of LLX steps is limited, since they must entail transfer to and from the solvent phase, under such driving forces as can be practically applied. The main steps always are extraction and stripping. All auxiliary steps are introduced around these two main steps. The auxiliary steps may be pre-extraction, solvent extract scrubbing, back extraction of product, each under the conditions which favor the step, provided the overall process technology and economics permit such conditions to be applied.

At this point a first "order of magnitude" evaluation can be made; if the approach is not shown to be unacceptable, then the next stage of experimentation can be planned. On the basis of cross current work and the auxiliary steps, a first flowscheme can be drawn up. If the system lends itself to graphical analysis, an equilibrium curve can be drawn for the selected conditions, the operating line entered as a function of material balance, desired yield, etc., and the number of transfer stages required thus obtained. The separation can then be tested in the laboratory in a multi-stage countercurrent simulation, proceeding until steady state is attained. Steady state here will be defined by analysis of exiting streams. It is safe to say that this steady state will be valid for major component(s); it cannot be assumed that steady state will necessarily have been reached with regard to minor components. One very important aspect is that the solvent entering the multi-stage system should be as near as can be anticipated to the composition of the recycle solvent which will be returning from the stripping step. In order to be able to anticipate the composition of recycle solvent it is desirable to perform a cross current stripping of extract, using the same philosophy as for the extraction. Again, on the basis of conjugate phase analysis, an equilibrium line and an operating line can be constructed so as to calculate the stages required for specific exit conditions at either end, i.e. the product or the recycle solvent. On the basis of this, the recycle solvent feed to extraction will be known, hence an adjusted operating line can be applied for the extraction graphical calculations and, therefore, for the countercurrent laboratory simulation. Once steady state is attained in the countercurrent multi-stage simulation, samples of conjugate phases should be analysed to verify the equilibrium curve for the main component, and in order to arrive at the distribution of minor components as a function of the level of the main component. If this distribution is strongly dependent on the main component, it may lead to internal cycling, hence to build-up of the level of minor component(s). To some extent the cross current data can show up this tendency, if the ratio of main component to minor component is seen to change as one goes along the cross current system.

If there is build-up it may take considerable effort to attain steady state in a simulation unless some special steps are taken with this in view. At steady state eventually what goes in will come out, but the levels in the system may be entirely different from those anticipated on the basis of feed composition only. In a countercurrent operation minor components can build up by orders of magnitude. If advantage can be taken of this fact by withdrawal of the required quantity of impurity at highest level of concentration this may be an efficacious procedure.

The countercurrent simulations will provide data for checking the flowscheme and the material balance arrived at from the preliminary testing. If significant corrections are required, then the preliminary economic evaluation must be rechecked too.

At this point it should be reasonably easy to outline a program of work for further development.

Multi-stage countercurrent simulations in the laboratory are usually the simplest way to check out the basics of a flowsheet proposal, but it is tedious work, particularly if intermediate steady state streams are to be generated for use in the testing of auxiliary steps. Probably less tedious is bench scale continuous testing, provided a reliable bench unit is available which can be put together in modular manner so that the process can be tested as completely as possible, incorporating all recycle streams. The extent to which the bench unit can be automated will be dependent on the degree to which it has already been run in a reliable manner. Nothing can be more inefficient and frustrating than attempting to bring a bench unit to the level of reliability and continuity desired while at the same time aiming at attaining some confidence level on the validity of a flowscheme proposal.

In general the cross current tests in conjunction with a multi-stage countercurrent simulation should give a valid flowscheme which can readily be verified on continuous bench scale unit. Extraneous phenomena, such as interfacial accumulations, changes in coalescence characteristics and secondary reactions in the solvent system, are more likely to be observed in a continuous bench unit, but even here there is much that will not be observed, often because of short test time, withdrawal of large sample volumes relative to the total volume in the circuit, long cycle time of solvent versus aqueous feed, etc.

The problem starts when there are a number of possible flowschemes, each with its own advantages and disadvantages. These cannot readily be tested in multi-stage simulations, mainly because production of valid intermediate solutions in sufficient quantity to permit feeding a next stage for a reasonable length of time may be a very long-drawn procedure.

Thus far, for complex multi-component systems very little use has been made of computer simulations, possibly because the data base required for valid simulations is not worth the work involved unless this is to be an ongoing operation. Thus for PD work it may seem to be of little aid. Once there is an operating plant generating data continuously the picture would be quite different and much more use could in fact be made of simulation in order to test possible variations and to evaluate the advantages they would offer, as well as to seek for build-up, etc.

Many other aspects can be learned from a bench scale unit: these are process control (as opposed to operating control), start-up procedures, shut-down, perturbations, etc. Temperature effects can probably not really be tested, but maybe they can be observed, and then carefully tested in experiments designed for these purposes.

When going out from a bench scale operation, it should be borne in mind that even though there is no need for the equipment to be a model of a larger unit of the same type, i.e. similitude is usually not required, nevertheless there is significance to the equipment type; certain conclusions regarding process stability, start-up, shut-down, control may be strongly interactive with the equipment type. Other types may react in a completely different way. Also, while a bench unit is not intended to be a model of the equipment, it is intended to provide a model of the process. Lack of similitude and the degree to which this can have process implications must be carefully considered.

Let us set up the following scenario—a number of processes are known for cleaning wet process phosphoric acid, and patents, reviews and papers have been published. Now suppose someone has a LLX system for cleaning up a different acid, using a solvent different from any of those describing operating plants for wet phosphoric acid clean up. The first question is whether this solvent which works for one acid will work for another, i.e. for phosphoric acid in this case. In principle one would say "presumably yes". The next question then is how does one set about getting the minimum of data in order to be able to foresee an R & D plan, *a priori* trying to place the stress on likely problem areas?

II. The LLX Separation—Chemical Feasibility

One approach to upgrading wet process phosphoric acid that has been published is based on upgrading only a part of the feed stream so as to obtain a clean stream and a dirty stream. Two approaches can be followed: (a) keep the dirty stream as near in quality to that of the feed, or (b) push the quantity of clean acid as far as possible, so that the quantity of dirty acid will be as small as possible, but at maximum impurity level. Furthermore, there are a number of impurities common to all wet process acids which must be reduced below some accepted maximum level in the clean acid, but there will also be specific impurities related to a particular wet process acid, its origin and history. These specific impurities may or may not fall into the general clean-up approach; in other words their reduction may or may not entail special steps. The approach of (a) means handling maximum wet acid for minimum clean product, that of (b) just the reverse. The choice between (a) and (b) will be in part economic, in part technological, in part local, depending on the specific case and circumstance. At first glance, therefore, the initial test work must cover both alternatives to permit evaluation in due time.

The information to be collected in the specific case is probably as follows:

1. Distribution coefficient of H_3PO_4 as a function of acid concentration.
2. Distribution coefficient of SO_4^{2-} and F^- as function of each anion concentration and the total acid level.
3. Distribution of cations commonly regarded as undesirable and also any specifically significant in this feed acid: here probably Fe, Mg, will be major cations; trace cations will be according to the known analysis of the potential feed, e.g. Zn, Cd, Cu, As, heavy metals (as Pb).
4. Solvent composition and limits of possible variation, in the light of known relationships to acid extraction.
5. Operating parameters, mainly probably temperature.

All of the above information can probably be derived by sets of cross current extraction. There is, however, a grave pitfall in such cross current testing—if water is transferred, the picture gained will be false since the relative total volume of solvent entering the cross current system is large, hence concentration of aqueous phase will take place due to water transfer to solvent. If this is so, it may be desirable in cross current testing to preload the solvent with the predetermined equilibrium level of water, so that essentially no water transfer takes place.

On the basis of the conjugate phases analysed the whole picture can be delineated, separation factors can be estimated as a function of the change in the main concentration variable, which is clearly acidity, also as a function of the solvent composition limits and the operating temperature limits.

As a general scenario, another interesting case is one where a metathetic salt–acid reaction is promoted by extraction of acid to move the reversible reaction in the desired direction. The specific case of KNO_3 production from KCl and HNO_3 has been described in principle in the literature. This example is particularly interesting since it makes use of a number of back and forth transfers to attain the ultimate aim of recovering both desired products from the metathetic reaction.

$$KCl + HNO_3 \rightleftharpoons KNO_3 + HCl$$
$$\text{Feed materials} \qquad \text{Products}$$

Let us now take another case, unspecified, and go through the steps which would be required to delineate the flowscheme for producing one salt from another by a similar metathetic salt–acid reaction, using LLX for moving the equilibrium in the desired direction by acid extraction. For the exercise we shall restrict ourselves to the use of a solvating solvent, such as a C_4 or C_5 aliphatic alcohol, as extractant.

For our purpose, we shall assume that a salt M^+A^- of monovalent cation and anion is to be converted into $M_2^+B^{2-}$ where B is divalent; this implies also that $M^+H^+B^{2-}$ is a possible product. The metathetic reaction therefore can be represented in two parts as:

$$MA + H_2B \rightleftharpoons MHB + HA$$
$$\underline{MA + MHB \rightleftharpoons M_2B + HA}$$
$$2MA + H_2B \rightleftharpoons M_2B + 2HA$$

The following is an initial scheme for a test program aimed at devising a likely flowscheme by identifying the likely regions of operating acidity:

1. General background about the aqueous system $M_2B/H^+/H_2O$, solubility and identity of solid phases as a function of H^+ level in a conjugate aqueous phase can most likely be obtained from the literature. Temperature should be considered as a possible variable.

2. Using the conjugate aqueous phases at the H^+ levels in equilibrium with the various solid phases possible in the system, determine the effect of adding MA, initially in the absence of the conjugate solid phases to identify solid phases if such precipitate, and in particular whether the H^+ level tends to drop; determine the limiting compositions when the solution becomes saturated also with MA. A good understanding of the aqueous system is very helpful when starting to examine the extraction system. The effect of temperature on solubilities and solid phase species should be determined.

3. Using selected aqueous phases, in the presence and/or in the absence of the various possible solid phases of the (M) (B) family, including also the case of MA saturation, determine the equilibrium alcohol phase under limiting conditions with selected aqueous compositions. Various aliphatic

II. The LLX Separation—Chemical Feasibility

alcohols, e.g. C_4–C_8, are probably useful; composition in organic phases should indicate $\overline{M^+}$, $\overline{H^+}$, $\overline{A^-}$, $\overline{B^{2-}}$, $\overline{H_2O}$ probably on wt/wt basis to avoid having to take volume changes into account. If the aqueous phases are expressed in "weight per unit of water", it may be convenient also to express the organic phase in "weight per unit of alcohol" since this will simplify subsequent comparisons and material balances.

The above information is substantially sufficient for selecting possible regions for operation. This can be done by noting the influences of parameters on compositions and expressing these influences in terms of product type, conversions and operating flowscheme. The KNO_3 scheme is performed conceptually in an invariant system, i.e. using an aqueous solution saturated with respect to both salts, KNO_3 and KCl, at fixed H^+ level and fixed temperature. In the case of H_2B acid, where two products MHB and M_2B are possible (or even three, if $MHB \cdot H_2B$ is known to exist at levels of acidity within the region considered practical for acid–alcohol–water systems), this adds operating considerations which do not exist in a system where acidic salts are not formed; these would have to be considered when selecting the best operating format, e.g. deciding whether to do a two-stage cross current type of conversion on solid, exactly as shown in the equations above, or whether a countercurrent extraction operation would be preferable.

Now the moment that some quantity of MA can be converted into an MB solid salt, the reaction has in fact proceeded to the right, but the problem of separating the equivalent quantity of HA remains to be settled. The quantity of H_2B accompanying HA into the organic phases at equilibrium with selected aqueous phases will indicate the extent to which this will present a problem in the separation and recovery of the HA co-produced in the metathetic reaction. The possibilities of separation within the LLX system itself—by analogy with a stripping/rectification distillation—need be examined (this is the manner used in the HNO_3–HCl separation). According to the nature of the two acids, and the relative quantities involved, other procedures can be considered, e.g. back reaction, evaporative separation, precipitation etc. The aim in all cases must be to return the alcohol solvent to the reaction system and to separate a stoichiometric quantity of A, but recovery of HA as such is not mandatory. If separation within the LLX system is to be considered, it is necessary to regard this now as an H^+, A^-, B^{2-}, H_2O, Alc system and to determine distributions and separation factors between two liquid phases in the normal manner.

The scenario described here serves to show that a well-thought-out minimum experimental program for determining limiting equilibria in multi-component, multi-phase systems will be sufficient to permit construction of the basic patterns for comparative flowschemes. Similar scenarios can be developed for other cases.

LLX Internal Recycle/Reflux — Extension of the Rectification/Stripping Analogy

In distillation the concept of reflux of condensate product to promote separation, by reverse transfer between the liquid/vapor phases, is well accepted. Similarly, stripping, by feeding pure vapor in countercurrent to the liquid feed, is equally standard. In order to apply these concepts to LLX, it is necessary to translate vapor/liquid into a liquid/liquid analogy; for convenience let us designate these liquids as "water phase" and "organic phase" respectively, eliminating the term "solvent" since either water or the organic liquid may be the extractant, according to whether one considers the rectification or stripping analogy.

It is obvious that these concepts apply to separation between at least two solutes, which have sufficiently different distribution coefficients to give a "separation factor" of accepted level to permit improved separation by rectification or stripping.

Whenever the separation is from an organic phase the reflux will be the purified aqueous product from the next step; whenever separation is from an aqueous phase the reflux will be purified organic product from the next step. This can be seen from the simple diagrammatic presentation in Fig. 26.

Another interesting analogy is with partial vapor pressure and the effect of a relatively non-volatile solute on the volatility of the volatile solute. Similarly in LLX a solute of lower distribution coefficient which does not distribute can cause increased transfer of the second component which does transfer.

In LLX usually, when using these analogies, first the material balance needs to be satisfied; the heat balance in some cases can be ignored since overall heat effects across the whole system may be small, even though this may not be so for each transfer step. Generally heat effects will be very pronounced in cases where ion-pair formation or complexation is the transfer mode, hence cooling or heating may need to be introduced.

In cases where a non-distributing component has an ion common with, or similar to, that of the distributing component, recycle can lead to a marked increase in concentration of the transferring component.

This can be seen from the simple diagram for acid extraction in the presence of salt as presented in Fig. 27; also from the extraction of Br^- from a high Cl^- brine, as can be seen from Fig. 28.

In LLX the recycle/reflux approach is not one that is generally utilized. It is in fact a very powerful separation tool and has been put to spectacular use in a limited number of special cases. Several examples will be presented here in simple block diagram form in Fig. 29, Examples 1 through 5.

Fig. 26 Distillation analogy

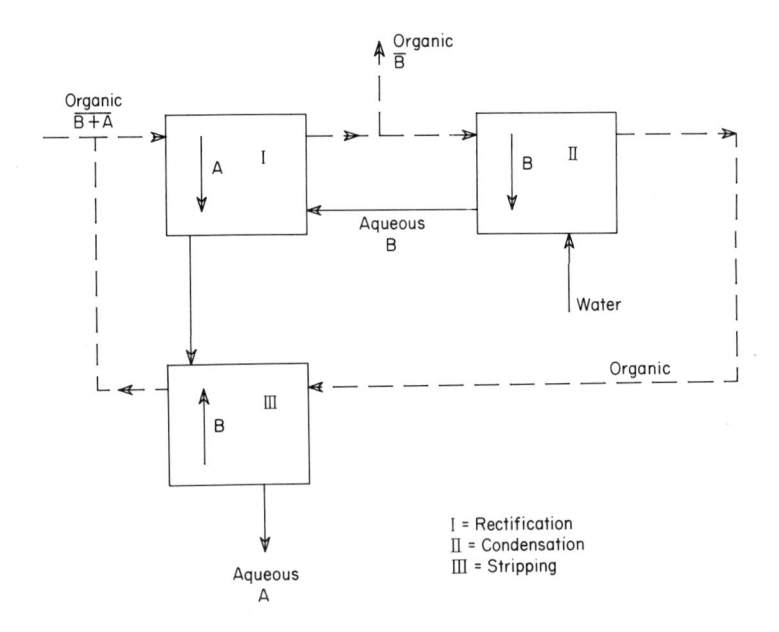

$$K_D B \frac{Org.}{Aq.} > K_D A \frac{Org.}{Aq.}$$

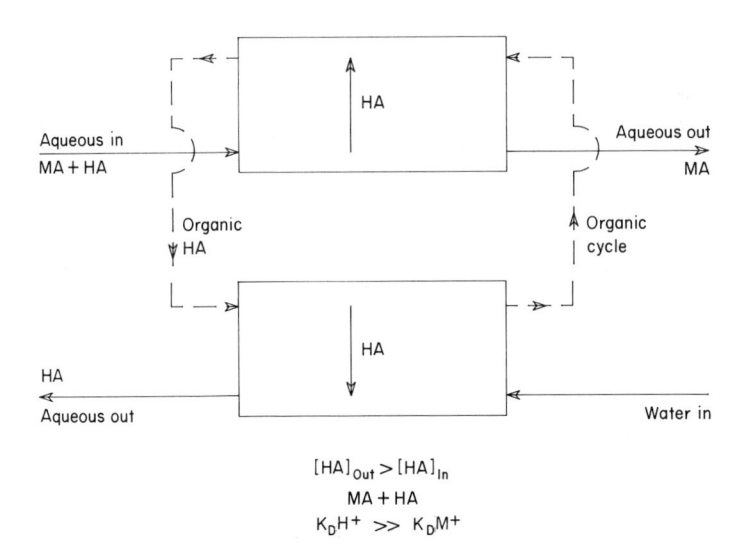

Fig. 27 Common ion effect

$[HA]_{Out} > [HA]_{In}$
MA + HA
$K_D H^+ \gg K_D M^+$

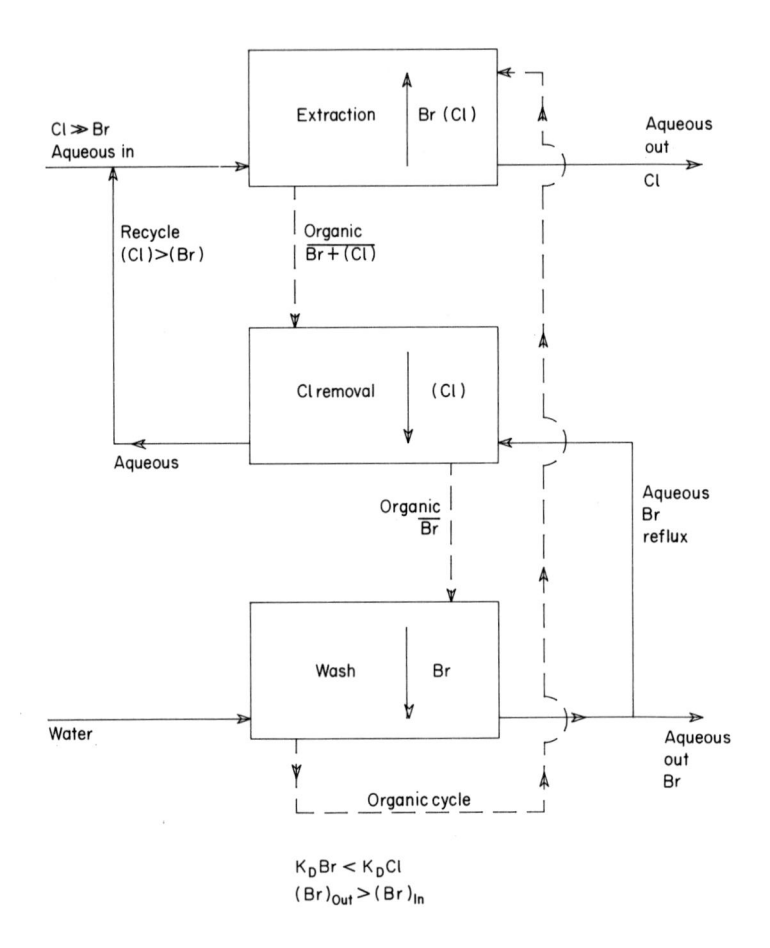

Fig. 28 Extraction of Br⁻ from Cl⁻ brine

Example 1
Pre-extraction of Fe^{3+}/H_3PO_4 process
U.S. Patent 3,497,330

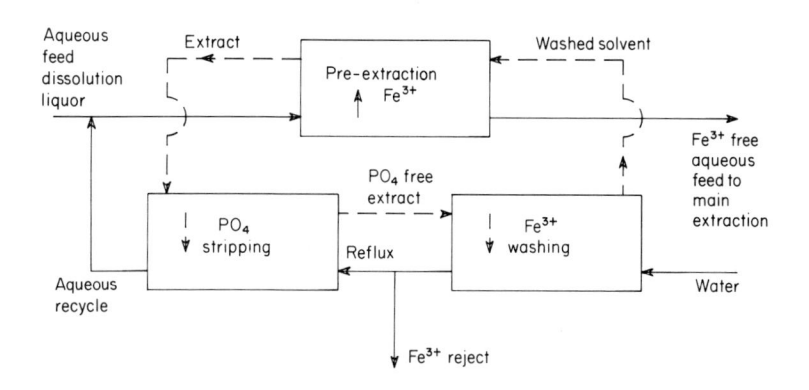

Example 2
Separation between acids HNO_3/HCl using pentanol

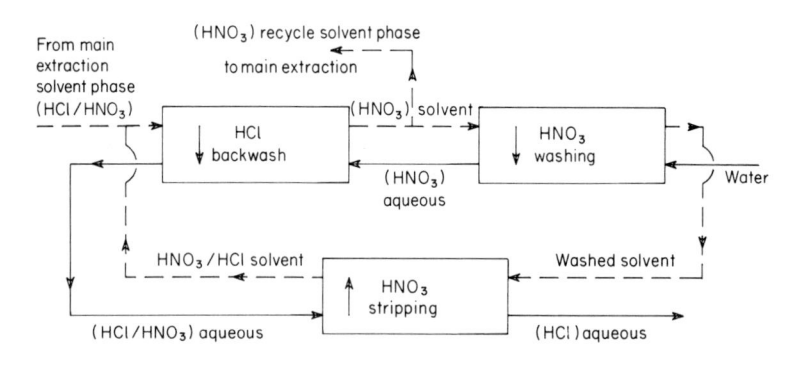

Fig. 29 Use of recycle/reflux as aids in separation
Example 1. Pre-extraction of Fe^{3+}/H_3PO_4 process
Example 2. Separation between acids HNO_3/HCl using pentanol

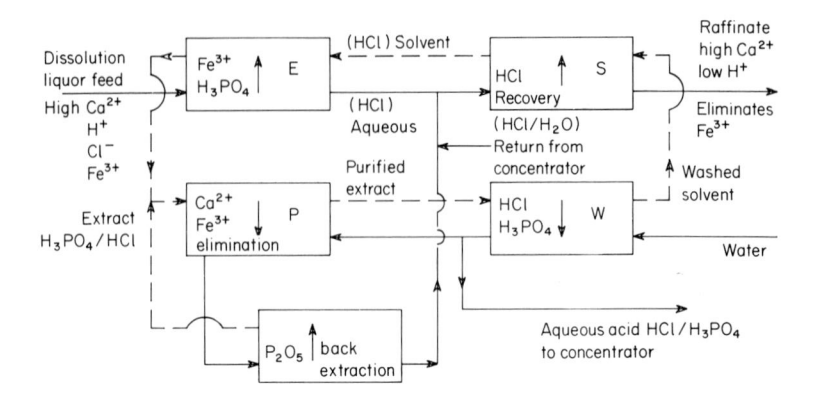

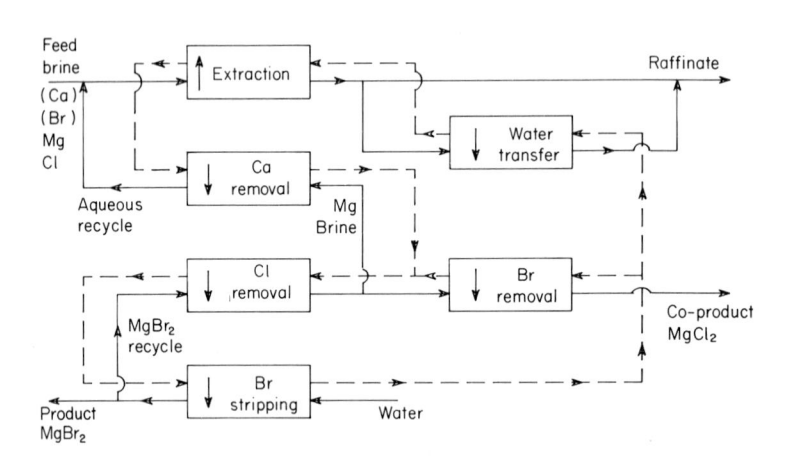

Example 3
LLX elimination of Fe^{3+} in H_3PO_4 process
by controlling Ca^{2+} and HCl

Example 4
Recovery of $MgBr_2$ and $MgCl_2$
from halide brines

Fig. 29 *contd.*
Example 3. LLX elimination of Fe^{3+} in H_3PO_4 process by controlling Ca^{2+} and HCl
Example 4. Recovery of $MgBr_2$ and $MgCl_2$ from halide brines

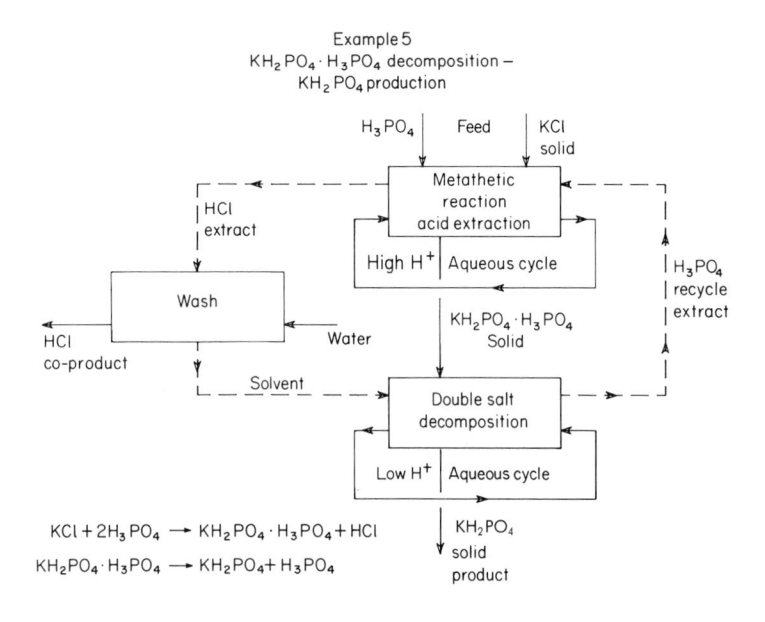

Example 5
$KH_2PO_4 \cdot H_3PO_4$ decomposition – KH_2PO_4 production

[Flow diagram showing:
- Feed inputs: H_3PO_4, KCl solid
- Metathetic reaction acid extraction box
- HCl extract → Wash box → HCl co-product
- Water and Solvent inputs to Wash
- High H⁺ Aqueous cycle
- $KH_2PO_4 \cdot H_3PO_4$ Solid
- H_3PO_4 recycle extract
- Double salt decomposition box
- Low H⁺ Aqueous cycle
- KH_2PO_4 solid product]

$KCl + 2H_3PO_4 \rightarrow KH_2PO_4 \cdot H_3PO_4 + HCl$

$KH_2PO_4 \cdot H_3PO_4 \rightarrow KH_2PO_4 + H_3PO_4$

Fig. 29 contd.
Example 5. $KH_2PO_4 \cdot H_3PO_4$ decomposition—KH_2PO_4 production

Example 1. Pre-extraction of components, minor in quantity but with higher distribution coefficients than the main component, using the same LLX system. The removal of Fe^{3+} from phosphoric acid produced by hydrochloric acid dissolution has been described in the patent literature, using a solvating solvent of alkyl alcohol type.

Example 2. Separation between two acids, using the extract phase as the feed to the separation system. This has been done in the KNO_3 metathetic reaction. It is an exact analogy to distillation, but using liquid–liquid in place of vapor–liquid equilibrium data to determine stripping and rectification. The separation is between HCl and HNO_3 in a pentanol solvent.

Example 3. Removal of Fe from phosphoric acid produced by hydrochloric acid dissolution using the extract as feed to a back wash purification stage followed by back extraction of the washing phase.

Example 4. Separation of bromide values from brine and concentration of the bromide by a type of recycle–rectification operation.

Example 5. Separation between an acid and its salts, e.g. in the case of multivalent acids like H_3PO_4. This has been done in the case of KH_2PO_4/H_3PO_4 where a double salt $KH_2PO_4 \cdot H_3PO_4$ may be the stable solid phase in certain cases.

All these cases relate to solvating systems which are fully reversible. In other systems, e.g. complexation, ion exchange, ion-pair formation, the same concepts can be applied, provided that transfer can be made reversible by some free manipulation. Often the manipulation variable will be the acidity or alkalinity level; in some cases a temperature difference may change the distribution sufficiently to make the transfer essentially reversible.

The fact that the examples relate to inorganic ionic systems is not meant to be limiting. Organic ionic and non-ionic systems can be used equally well, applying the basic concept of the relative distribution coefficients and reversibility.

Computer Simulations in LLX

The building of a flowscheme for achieving desired separations follows directly from two things, the limited possible types of back and forth transfer in LLX systems, and the simplicity of the approach of applying the concepts of limiting conditions and invariant systems. This means that valid flowschemes can be devised with the minimum of experimental equilibrium data.

The building of a model is in a sense just the reverse. A considerable data bank is required in order to arrive at a valid correlation, which can then be used for predicting the effect of proposed changes, provided one sticks to interpolation and not extrapolation, or at most very limited extrapolation.

These two basic differences imply that the two areas do not overlap, are not mutually exclusive, but more likely mutually supportive. What this means is that a model cannot lead to a process but once the process has been defined the model will be very useful and will save considerable work, in plant operation, optimization, and control. This applies particularly to separations between similar components. If the model is based on a good correlation of equilibrium data it is probably very simple to find the effect of splitting the flows, increasing or decreasing volume ratios, returning streams as recycle, increasing the number of theoretical stages, changing feed compositions, achieving changes in exiting stream composition, etc. Possibly also, the critical control point can be identified, e.g. the point most sensitive to small changes in feed streams, whether these changes relate to quantity or to quality. Furthermore, to the extent that various back and forth transfer steps have been included, the effect of changing the order or eliminating steps, or probably including steps of the same type, can be rapidly and usefully analysed by such a model. It seems that the model cannot offer help in conceiving a completely different approach even though the change envisaged may be acceptable in the general concepts of LLX.

A model derived for one system can be expected to be useful for a sister system provided the limiting controlling concepts are common to both. Thus an amine/acid system which has been modelled for one amine can probably be applied to a different amine provided a correlation can be included connecting the amines—such a connection may, for example, be the half-titration pH at fixed temperature using the acid in question. Similarly, going from one acid to another with the same amine by using the pK's of the acids may be possible. This may, however, not be so simple if some other equilibrium constant is controlling, e.g. solubility of the amine–acid ion-pair in the composite non-aqueous (solvent) phase, or the hydrolysis constant, etc. The next question would be whether the diluent can be changed for a specific amine/acid system, provided some correlation relating the diluents can be included.

LLX–Process Synthesis–Computer-Aided Process Design

While LLX is analogous to distillation and in classical chemical engineering terms both were classed as unit operations, yet in modern LLX there are many aspects in which LLX differs in principle from distillation. Basically both are dependent on two-phase equilibria in which distribution of each component between the phases is uniquely defined by some coefficient of thermodynamic activity. While this is always true it is possible in LLX to choose the second phase so that the quantity of this component that passes into the second phase may differ greatly from case to case.

A comparison of water transfer from a specific aqueous phase to various solvents which themselves have acceptably low solubility in the aqueous phase will in fact show this clearly, since the activity of water in the several conjugated aqueous phases will essentially be the same. A solution of a salt which has very limited solubility in the solvent phases being examined is probably a good stable aqueous phase for comparison, preferably at salt saturation. This has been discussed in the framework of a conjugate system for determining partial vapor pressure of water for brine solutions, or for concentrating aqueous solutions by water extraction. The point is that entirely different levels of water in various solvents will be in equilibrium with essentially the same aqueous phase, in other words, they are all essentially at equal activity in regard to water. Depending on the purpose of the LLX contemplated this may be an important aspect, or an irrelevant fact, or may simply serve as a scale for indicating relative hydrophobic characteristics of solvents.

If one accepts the approach that distinguished between what was called a unit operation and a unit process, then LLX is close to a process in which an equilibrium reaction is promoted in a desired direction. This means that LLX can in fact be generalized into the very wide range of chemical transport, but this has not usually been recognized, hence LLX in practice has become compartmentalized instead of generalized.

Let us now look at computer-aided process design (CAPD) for LLX, accepting the fact that greatest strides in this area have been made in fact in distillation, and therefore the tendency will be to follow the guidelines of the latter. Now this is not a negative approach at all for each special LLX system, since in a defined two-phase system the analogy to distillation still remains great. However, if we accept the fact that the choice of LLX systems for a specific purpose is wide, then it is legitimate to ask whether the CAPD can help in selecting the best choices. It would almost seem as if network design is a closer analogy conceptually, since it is a mode of selection. Certainly the concept of "reduced states", no matter how these are defined, has been found to be a very powerful tool for making comparisons and selections in LLX, especially if this brings us to some concept of invariant or pseudo-invariant system, i.e. if it helps to eliminate degrees of freedom and thus to limit or define the choices.

This concept of "reduced states" is in fact clearly used in comparing driving forces in the classical procedures used for calculating transfer stages for distillation or LLX, where reference is made always to the equilibrium distribution value for each particular concentration. For second-phase selection this concept of "reduced states" can be somewhat differently applied; by selecting an invariant system (e.g. a saturated salt solution in the presence of solid phase at fixed temperature and pressure), the distribution levels of solute in various second phases will be absolutely comparable for evaluation, when at equilibrium within this invariant system.

For the skilled process developer, therefore, the task is to arrive at some type of classification of systems which, while encompassing the overall broad area of choice, would nevertheless reduce the individual choices to a limited number of realistic groups so that the process designer can take over and aim at arriving at best combinations or at identifying limiting or forbidden combinations.

Fundamentally, the classification for LLX must depend on chemical interactions since it is these that extend the scope of choice. The main interaction will be between the extractant and the solute, but secondary important interactions will be between the modifier and/or diluent and the solute.

Recently computer programming has been applied as an aid in design or control in LLX. To the degree that a case has been extended to cover also the process design or even the process optimization, this can serve as a starting point for developing a generalized approach.

Process Synthesis (CAPD) Generalized Models—For What?

In LLX, once a flowsheet has been laid down and once equilibrium data and material balances are available so that distribution coefficients and separation factors can be calculated, it is possible to derive a model for each transfer operation and to connect them so as to represent the case as a whole. It is then possible to use this model as an instrument of control, in two ways, namely for anticipating drifts from steady state—as would be shown, for example, by changes in concentration of critical components at selected points—and for correcting the drift according to plan, or for anticipating the effect of changes in feed composition and thus to specify appropriate changes in operation so as to obviate variations in quality, concentration or recovery yield of product.

It is also possible to use the model as an instrument of optimization for the specific case, thus recycle can be optimized from various points of view, e.g. as a function of quantity, composition, point of generation, point of return, etc. Another aspect of optimization is the calculation of likely savings by installation of an additional piece of equipment and pointing to the best position for such installation.

These and similar uses make it worthwhile to have such a model for any defined LLX process which has been designed for attaining specified targets; this says that only *after* process synthesis does one come to computer-aided process design and operation.

Now before the question arises as to whether we can arrive at computer-aided process synthesis itself an earlier question is whether the process model can, in principle, aid in selecting from among equipment types. This would require the limits of equipment types to be specified as a function of system parameters, even if only the three reference systems of the EFCE are used for this specification. The process model would then define the process limits which could then be imposed on the equipment limits. This would probably be very useful in cases of extreme variation in properties or quantities as a function of process, as it would show whether optimization of equipment choice leads to an equipment mix, or whether one selected equipment type can cover the whole range of process.

As regards process synthesis models, a simple case would relate to an existing scheme with a specified second liquid phase, when for some reason it is desired to convert to a different specified second phase, while keeping to the format as far as possible, and then subsequently optimizing for the new second liquid phase. Presumably modification of the existing model, provided one knows how to introduce true distribution coefficients and true separation factors, will lead to a modified model for the new case which can presumably then be checked out and used for operation control and optimization in the manner discussed above.

The next question is whether one could define the characteristics of an ideal second liquid phase, in other words, specify desirable distribution coefficients or limits of distribution coefficients, or limits of separation factors, and then show how best to fit the separation steps together, assuming that one can utilize only pre-extraction and extraction on the aqueous feed line, and back wash or scrub, and strip on both separate extracts or second liquid phases; also that recycle of such scrub is required in both cases. This would of course require that mode of transfer be independently specified, as this will give the limits of transfer as a function of concentration of solute, and concentration of transfer agent if such is to be used.

Since the steps in LLX are in fact repetitive and limited, consisting always in transfer to and fro between the two liquid phases under specified flow ratios and under isothermal or adiabatic conditions according to the process, it should be possible to set up a general model with all parameters designated for both phases being fed to a transfer system. There is a problem of course in cases where considerable change in properties occurs as a function of the mass transfer, but probably this can be overcome by describing a "rich" and a "lean" regime as far as the transferring material is concerned. If such overall models can be set up this would go a long way in aiding process synthesis.

Transfer Modes

Very few case histories of LLX processes and their implementation have been published. Those that have been described relate in the main to the basic aspects of implementing the main separation of the process, in other words only the principal transfer mode is considered. Possibly the reason for this is that almost all LLX processes describe separations among similar components, e.g. cations, anions. In such cases a scale of distribution coefficients can be postulated as a function of some process variable. Classical cases will be separation of cations as a function of pH, using an acidic extractant, or the use of a complexing agent with considerable specificity, e.g. oximes, etc., or the use of an acid to complex particular cations, such as:

$$HCl-HFeCl_4 \text{ vs. } H_3AlCl_6$$

The problem may be different in the case of separations for biotechnological purposes when the nature of the organic extracting phase changes by virtue of the main transfer itself, hence there is the possibility of simultaneously having various modes of transfer. In process solutions where various materials may be present, possibly in orders of magnitude differing concentrations, there is every reason to relate to such differing transfer modes. It is true, of course, that modes of transfer need not be taken into account when describing a system, since the only aspect that is necessary for describing a system is the *distribution coefficient* of each component under specified equilibrium conditions. There is a tendency, however, to ignore those components which cannot transfer by the same mechanism as the main transfer. When there is nevertheless transfer of such components, it may be necessary to reconsider the system characteristics so as to postulate modes of transfer for secondary components. Such a postulate will then help in arriving at an approach for separating such secondary components, if the main objective is impaired by their accompanying the separated main component.

Let us in principle therefore define three modes of transfer: first, positive transfer by the main procedure selected for the main component; second, "carry over" by a secondary procedure dependent or independent of the main transfer; and third, entrainment to whatever degree this occurs. Often it is difficult to distinguish between the second and the third procedures. The main distinction must be the degree of preferential transfer of components— entrainment implies *no* preferential transfer, i.e. phase A as a whole is entrained, irrespective of the actual quantity; this means that the ratio of components in A will be preserved also in the phase B apart from the positively transferred material. Carry over implies preferential transfer by a secondary mechanism. Since it is unlikely that all components of phase A will transfer equally by secondary mechanisms, one can expect changes in the ratio of such components, even drastic changes. The cause of such preferential transfer may be the nature of specific components themselves, or the character of the conjugate phase, which may even be changing by virtue of the main transfer itself.

Procedures for handling the problems of minor transfer, whether this be by positive transfer, or by carry over, or by entrainment, may in fact not be different, since the type of steps possible in LLX are rather limited. All modes of undesirable transfer can be counteracted by interposing a step such as a partial strip, wash, or scrub. The efficacy of the step will depend upon the case in point.

Entrainment will essentially be the result of the equipment used for generating surface for mass transfer by dispersing one phase in the other and then permitting the dispersed phase to coalesce. The characteristics of the system itself will have a bearing on the ease of dispersion and of coalescence in relation to a specific piece of equipment, so that even entrainment will be an interactive phenomenon.

Entrainment will be counteracted by the interposed step to the extent that the equipment employed for the scrub does not re-entrain, and depending on the composition of the scrub conjugate phases.

Extraction of minor components by the same mechanism as the positive transfer can be countered by a scrub step directly related to the relative distribution coefficients of the components. Only substances of lower distribution coefficients than the main component can be eliminated by such a procedure. Material of higher distribution coefficient must be pre-extracted.

Transfer by carry over implies an undefined mechanism depending upon the case, hence it is not possible *a priori* to predict the efficiency of scrub in such cases.

As an exercise, let us assume that carry over is by partition, directly related to concentration by a straight line relationship. Normally we will be talking about a partition coefficient way below the distribution coefficient of the primary component by positive transfer. In a sense then the same scrub concept can be utilized as proposed for minor components of lower distribution coefficient. However, process-wise this may not necessarily perform to the same degree, since the shape of the equilibrium lines may be different. A material balance and a stage analysis will show which is determining, i.e. whether the scrub step should be designed by using as a yardstick the elimination of minor components transferring by the positive transfer mode, or using elimination of carry over components as a yardstick. In either case, a straightforward study using separation factors can give the optimum scrub operation.

Transfer of feed components by entrainment is a different story. First of all this is kinetically controlled and not equilibrium controlled; secondly it is interactive with the LLX equipment being used and the conditions selected for the particular mass transfer steps. Entrainment can be controlled by determining the dispersion type desirable in a particular step and then attempting to satisfy this within the process limits.

Thus dispersion and coalescence are highly interactive; mass transfer rates too may be dispersion type related, depending on mass transfer mechanism, solubilities, etc. If the equipment used for scrub and for extraction is of similar type, then taking into account flows, etc., the entrainment will be similar; hence contamination of product will in the final instance be a function of the scrub conjugate phase at the solvent outlet. The more stages or the steeper the profile the better. It is probably necessary to distinguish between *primary* entrainment resulting from the droplet size distribution (e.g. in a dispersion of water-in-oil, the smallest aqueous droplets may not coalesce and may follow the solvent phase) and *secondary* entrainment due to flow dynamics.

In some complex systems it may not be easy to isolate the causes for the appearance of some level of impurification in the main product. Very carefully planned experimentation and well developed analytical procedures may be required in order to arrive at a clear-cut picture. Not always is such a study justified—instead it may be possible to postulate transfer and try to arrive at the cure, then to test the cure—if it works that is all that is necessary. Still, if two postulates require different cures, the proof of cure may confirm the postulate. However, if two postulates are satisfied by the same cure, success does not spell understanding.

Typical Examples

In recent years a large number of processes have been developed for upgrading wet process phosphoric acid. A number of the processes have been patented and/or described in the technical literature; however, many are "in house" developments, and details are not known. This is, therefore, a good case to select as an example of an R & D project, since there is much information in the public domain to serve as analogy, and the field is still open for new developments.

The philosophy for upgrading wet process phosphoric acid has been discussed previously; the impurities present are known, the required specifications are known. The clean-up can be handled in a variety of ways. One consists of separating a smaller fraction of pure acid, leaving a larger fraction somewhat less pure than the initial feed; this means using a large volume initially. Alternatively high recovery can be sought, leaving the minimum of phosphoric acid to accompany the waste impurities.

Let us now follow through on the R & D study. For extraction of acids either solvating type solvents or basic solvents can be used. Both classes have been cited for recovery of phosphoric acid; the former includes alcohols, ketones, esters, ethers, amides, trialkyl phosphates; the latter mainly alkyl amines.

Let us go through an exercise of testing solvents to evaluate their applicability for the purpose in hand. First of all, we expect of the solvent: availability, stability, relative cheapness, acceptable degree of immiscibility

immiscibility with water/acid systems, selectivity to some degree, capacity. All these aspects need to be quantified. A wet process phosphoric acid is too complex a mixture to warrant using synthetic solutions in the study. On the other hand if the initial work is to be kept to a reasonable minimum it is not possible nor advisable to study the distribution of every component of the wet acid, but rather to select a number of important ones and regard them as being indicative of what can be expected. Thus sulfate and fluoride are main anionic impurities, while Fe, Mg and Ca are undesirable cations from the point of view of product. The following are basic questions in evaluating a solvent. What is the distribution of the main component as a function of concentration and temperature? What is the influence of impurities on this distribution, and vice versa? What is the separation factor between the main component and the impurities? Is the transfer reversible? Is there water transfer between aqueous and solvent phases? What is the maximum concentration that can be obtained from a specific feed? What is the degree of extraction that can be attained? Does the product or the raffinate control the ratio of solvent to aqueous phase, for a specific degree of extraction?

In a complex system the best way to obtain distribution data at maximum concentration is by determining the equilibrium at "limiting conditions" of feed composition and concentrations. This will give the distribution of each component, at its maximum feed concentration, in the environment and at the concentrations of all the other components. These distributions can then be used for calculating distribution coefficients and separation factors.

The next step is to determine the distribution of the major component(s) as a function of concentration. The simplest way to achieve this is by "cross current" extraction, as this allows the natural interactions to have full play. Since all components will distribute to some degree, it is possible by this method to obtain the effect of concentration and interaction on the distribution of minor components. Equilibrium curves can be constructed on the basis of the data, using concentration or ratios for the various components. From this the number of theoretical stages required for transfer can be determined. Strictly only a stage-to-stage construction or an overall mathematical model can be used for really complicated systems, but simplifying assumptions can be made in order to apply standard graphical procedures. Thus an adaptation of the McCabe-Thiele procedure for graphical presentation of a countercurrent system can be used, provided a straight operating line can be postulated. Various examples of such an approach have been reported.

In a complex system in which minor components distribute as a function of the concentration of some major component there is the danger of build-up within a multi-stage countercurrent battery. (Battery is used here to mean a physical assembly of transfer stages, irrespective of the actual equipment used.) This build-up can occur when conditions at the two ends of the battery are such that mass transfer of a minor component is reversed, meaning that the exit is blocked and that the component remains inside the system and accumulates there until its concentration is high enough for breakthrough, so that steady state is then attained.

Distribution coefficients and separation factors, as well as equilibrium distribution curves for the operating ranges envisaged, can be compared for various solvent systems and the apparent efficacy of various solvents evaluated. (The true efficacy of a solvent for a process as a whole cannot, however, be evaluated only on basic aspects of selectivity and distribution, since a process is a complex concept with many interactions.)

There are certain constraints imposed by virtue of the fact that the topic under consideration is LLX, which implies two things at least: first, that something needs to be separated, and second, that there will be transfer between two liquid phases. At the present point we assume that we know the purpose of the exercise, i.e. we know what we wish to attain; we also assume we know what must be separated from what but not necessarily which will be the one to extract and which to leave behind. The next thing we need to determine is the nature of the materials to be separated, whether they are similar or dissimilar, and which property of each can be utilized to promote the transfer of one away from the other. At this point we can also define the nature of the counterpart phase into which the transfer will take place in either case, by virtue of the property selected in either case, as the basis for mode of transfer.

Let us take three cases of the commoner types and analyse them along the lines specified above!

(a) Leach liquor containing metal values, desired and undesired say A and B; counter ion SO_4, due to leach acid H_2SO_4; liquor mildly acidic, pH $\simeq$ 2.
(b) Salt and acid of the same anion, i.e. MX and HX.
(c) Fermentation product, low concentration, impurified by the presence of inorganic salts and residual substrate.

We can now define the three cases as follows:

(a) Cations to be separated from the common anion. Two procedures are possible, selective or specific complexation (if a suitable reagent is known) or preferential cation exchange as a function of pH.
(b) Acid extraction, by ion-pair formation or by using a solvating solvent.
(c) Here the case is less clear-cut. Questions need to be asked and answered. Is the fermentation product acidic or basic? How is the product usually recovered? (If not this product, then some similar product, to serve as analogy). In either case will ion-pair formation be the route? If so, what will be the diluent/modifier choice to obviate partition of substrate contaminants, and so on.

In each of the three example cases much can be derived by comparing with case histories, well documented in literature.

Case (a). Selective, specific complexation of copper is a topic that has received widespread attention. A large number of variations of basic types of essentially specific complexants have been synthesized, each to answer a defined aspect in operation. Thus a recent reagent is for extracting at low Cu concentration in the presence of high Fe^{3+}. The reagent *proposed* selects the Cu and rejects the Fe^{3+}. If case (a) refers to Cu and Fe the approach in the first place is clear-cut; a flowscheme can easily be outlined on paper, data for the material balance can readily be obtained by a minimum of tests, and a first process evaluation can be made. When Cu is the desired metal value to be recovered it is very unlikely that preferential cation exchange as a function of pH would be considered at the present time, although this might well have been the case two decades ago. However, if Cu is not the desired metal value in case (a) the situation is far less satisfactory. If it is correct to say that the basis of the synthetic program for copper-extracting reagents was derived from the specific reagents of analytical chemistry, then although there are specific reagents for other metals, apparently there has not been the same drive to do a similar series of syntheses (for Cu the requirements of the reagents were well specified, perhaps other cases are less well specified).

Let us assume that Zn is the desired metal value to be separated. At present the best reagent is probably a phosphate ester which has the ability to separate Zn at the highest acidity, i.e. lowest pH; when acid leach is being examined, this will always be an important aspect. Again, there are sufficient examples in the literature to enable a reasonable flowscheme to be postulated as the basis for the minimum of essential testing so that a material balance can be derived and a preliminary operating flowsheet outlined, both for first evaluation.

Let us now go to case (b), separation of salt and acid of the same anion, i.e. MX and HX. At first glance one assumes that it will be preferable to extract the acid. One of two procedures comes to mind immediately. The one is to extract acid by ion-pair formation, the other by using a solvating solvent. In both cases, the presence of the salt with common anion is beneficial since it affects the ionization equilibrium of the acid in the desired direction. It is clear that the problem will be decided by the stripping step, not by the extraction step at all. In the case of the solvating solvent, the stripping is a fully reversible transfer by virtue of concentration—hence the benefit from the presence of the common ion in extraction is in fact seen in the stripping. In ion-pair formation the strip will or will not be reversible, depending upon the acid in question. The parameter change to promote back transfer will be a raise in temperature. The weaker the acid, the better the transfer, as a function of temperature rise. With stronger acids, a competing cation must be introduced—in certain cases this may be a viable approach, in others not.

Case (c) is quite different. At first glance the easiest way to select a procedure will be based on the current separation mode for the material under consideration, since probably some property is already being exploited for recovery purposes. Let us assume the material is currently being separated by cation exchange. Clearly there will be competition by the second cation for the anion of the reagent, while there is also the competition between the counter ion and the reagent for the cations. The ionization constants for the two cases will determine the situation, as well as the relevant molar concentrations. Undoubtedly some special step will have to be taken to arrive at a reasonable solution.

$$\overline{RH} + O^+X^- \rightleftharpoons \overline{RO} + HX$$
$$\overline{RH} + I^+X^- \rightleftharpoons \overline{RI} + HX$$

Reactions in LL Systems

The important aspect which differentiates a two-liquid phase system from a single-liquid phase is that there is separation; since the two-liquid phases are different, they introduce an extra degree of freedom. This may be obvious, within the framework of the Gibbs Phase Rule, but it is worth while stating positively.

Another obvious aspect is that at equilibrium the activities of components in the two phases are equal, even though absolute concentrations are not. For the practical application of two-liquid phases in reactions, this is undoubtedly the key fact. A simple example is the complexation of an ion in one phase, by an agent which has very low solubility in the second phase; the distribution of the ion will thus be increased in the direction of the complexing agent, so that its concentration in the complexed form may be much higher than it would be without complexation. However, the activity in the two phases at equilibrium will of course be equal, and it is this that is exploited to promote reactions in LL systems.

Thus when an acidic component soluble in an organic phase is used as an ion exchanger to extract a cation from the aqueous into the solvent phase, this is in fact a metathetic reaction which exploits the presence of two phases and the differences of solubility in two phases. The following two equilibrium equations show this very simply:

$$MA + HR \stackrel{K_1}{\rightleftharpoons} MR + HA$$

When only one phase is present, and provided solubility products are not exceeded, equilibrium according to K_1 will be established, but nothing practical will have been achieved. Now when for instance we take the case of HR which has very low solubility in the aqueous phase, we can set up a similar equation

$$MA + \overline{HR} \underset{}{\overset{K_2}{\rightleftharpoons}} \overline{MR} + HA$$

but with the difference that the bar represents solvent phase, and separation between M and A has been achieved. This example is so accepted a case that metal extraction which exploits this is rarely even described as reaction in a liquid–liquid system.

The case of promotion of a metathetic reaction by extraction of acid by a solvating solvent is well documented and well established in technology, especially in the case of the conversion of KCl to KNO_3 according to the reaction

$$KCl + HNO_3 \rightleftharpoons KNO_3 + HCl$$

The aqueous phase may be represented by the smallest physical quantity, provided that equilibrium has been established between the two liquid phases. Furthermore if both solid salts are present the system becomes invariant at fixed temperature and H^+ levels, hence in fact the reaction is promoted to the right by virtue of HCl transfer into the solvent phase, since K^+ transfer is very low. Recently something quite similar has been shown for promoting enzymatic catalysed equilibrium reactions in systems where product is removed from the aqueous phase which contains the enzyme, and hence where the reaction proceeds by virtue of distribution to the solvent phase.

Phase transfer catalysis also constitutes reaction in two-phase systems. Typically, one or more of the reactants are organic liquids, or solids dissolved in a non-polar organic solvent, while the co-reactants are salts or alkali metal hydroxides in aqueous solution. Phase transfer catalysis is the technique by which reactions between substances located in different phases are brought about or accelerated.

In the cases mentioned above the reaction mechanisms may be different, but they are all reactions in LL systems.

Reactions in multi-phase systems can be performed in multiple equilibrium stages if this promotes the desired reaction, leading usually directly to the classical countercurrent system. However, cross current contacting or single-stage operation may be preferred modes selected in specific cases.

A simple type of reacting system, for example, is the case of extraction of acid from some impure ambient phase and then reaction of the extract with a base to produce a desired salt. Thus a fermentation system may produce an acid which can be extracted selectively by some solvent, leaving the substrate residuals and metabolites in the aqueous raffinate; treatment of the extract with selected aqueous bases at determined pH and temperature will produce aqueous solutions of salts; even the solid salt itself may be obtained directly if the composition and quantity of the conjugate aqueous phase is controlled.

Methathetic reactions in inorganic or mixed ionic systems are established in liquid–liquid systems. Equally one can consider hydrolytic reactions as shown below, by extracting acid.

$$MA + H_2O \rightarrow MOH + HA$$

$$2MA + H_2O \rightarrow M_2O + 2HA$$

Here, too, precipitation of solid product may be attained.

There are many other reactions which could benefit from being carried out in two-liquid phase systems. Thus one can conceive of cases where reactant(s) and product(s) have different solubility characteristics, or different vapor pressures, or even different surface tension characteristics, so that the two-liquid phases may promote reactions or separations, as the case may be.

Still another example may depend on difference in solubility of a reactant gas in a different conjugate liquid phase. Since mass balance eventually describes the extent of conversion, a higher transfer per unit of volume due to higher solubility may have a far-reaching effect.

Accent has recently been placed on separations in dissociating systems, where differences in dissociation constants can preferentially promote transfer to the second liquid phase. The use of differences in formation or stability constants is probably the best exploited in liquid–liquid systems. All these reactions are promoted by the fact of the separation itself, which is inherent in having two liquid phases in the system.

Acid extraction, or perhaps anion extraction and hence also the extraction of salts in certain contexts, follows two different routes. When weakly basic solvating solvents are used, the acid extraction is strictly concentration controlled. The higher the concentration the higher the acid transfer, until the plait point is reached, then extraction ceases since there is only one liquid phase present. Examples of solvating extractants are alcohols, ketones, ethers, esters. Their distribution diagrams are all completely similar for inorganic acids, some quantity of water accompanying the acid extracted, although the extractant itself can partly provide the solvation for the acid; usually the water accompanying the anion and/or the cation on a molar basis will be less than the molar ratio in the aqueous phase; in other words on a solvent-free basis the acid is more concentrated with respect to water in the solvent than in the conjugate aqueous phase.

With stronger bases such as alkyl amines, which become protonated and give ion-pairs with the anions present, the limit is essentially as in any neutralization, depending on base strength of the amine and valence and strength of the acid, i.e. on the relative pK's of the acid and the base. This type of extraction resembles a normal acid–base titration. Usually there will be very little water accompanying the acid into the solvent, probably not more than mole per mole.

Amides have the characteristics of both types of extractant, consecutively. Thus at low acid concentration the effect is essentially like that with amines, while at higher acid concentration it becomes typically a solvation type extraction, strongly dependent on concentration of acid in the aqueous phase.

Many interaction types can be exploited to promote transfer back and forth between the two liquid phases. These interactions may take place in the aqueous-rich phase or in the organic-rich phase, and may be directed towards the extraction or the stripping step. The simplest case is the common-ion effect in the aqueous phase that promotes acid transfer to the organic phase. Another simple concept is the addition of a non-polar diluent to an extract to change the polarity of the organic phase and thus to cause transfer back from the organic to the aqueous phase. Both the procedures above have an effect similar to that of a concentration change as driving force for transfer, hence these procedures are applicable to the case of weakly basic solvating solvents.

With strongly basic solvents such as amines, a different approach must be used. Again polarity of the organic phase can be manipulated by utilizing diluents and/or modifiers judiciously. Alternatively, extraction and stripping may be performed at different temperatures which affects the pK's of the system. Still another procedure is to use a competing acid of selected pK which is essentially not soluble in the aqueous phase—this therefore has a strongly modifying effect on the overall acid–base system, and in conjunction with temperature will influence extraction and stripping in the desired direction.

The distribution of mixed acids in a solvating system, pragmatically, will be the resultant of the distribution of H^+, of the separation factor of the anions, and in some sense of the effect of the presence of the second acid on the first.

Thus in the system H_2SO_4–HCl–nBuOH–H_2O the system can be presented as a family of ternary curves with the sum (H_2SO_4 + HCl) as one of the apices. Each curve represents a definite H_2SO_4/HCl ratio with all the curves lying between the two ternaries H_2SO_4–nBuOH–H_2O and HCl–nBuOH–H_2O.

Metal and Acid Extraction

A monograph at research level should stimulate research by showing what has been accomplished and how this was done, and also by indicating what has not yet been achieved.

There are various general areas which have received major attention in LLX and where there are real success stories. One of these relates to cation separation and recovery (usually inorganic cations within the framework of hydrometallurgy), another aims at anion separation and recovery either in the form of acids or salts.

In metal extraction there have been major developments in a few cases during the past two to three decades, but each successive review article during this period continues to look forward to a widespread breakthrough, which does not come. As late as 1980 it was still only possible to point to uranium and copper, with cobalt and nickel interesting but not quite making the grade. One can speculate about the reasons for this situation but no single reason can explain, nor can a single line of reasoning predict.

The types of solvent which can transfer metal ions from an aqueous to an organic phase are limited to chelating, ion-pair forming, neutral or solvating, and mixed, i.e. acidic extractants which show both chelating and solvating characteristics.

If one looks at unit operations for separation, other than LLX, one sees that some do not need the intervention of an agent to promote separation since they rely on intrinsic characteristics of the material being separated; thus distillation, crystallization and sometimes precipitation are of this type, while extraction, absorption and adsorption need the mediation of an agent for achieving separation. In adsorption the trend has been towards developing a limited number of types of adsorbents for broad applications, while in extraction the tendency has largely been towards specific extractants for particular cases. The reason for this basic difference in approach may lie in the difficulty in identifying or even defining materials to be adsorbed (e.g. color, odor), while materials to be extracted are usually identifiable and are therefore handled as named compounds.

Without going into the pros and cons of these approaches, one sees that indeed the format of applications of LLX usually derives directly from the specific approach. Little has been done to generate a continuum of solvent systems from which to select for specific cases.

If one accepts the concept that LLX is equilibrium controlled (only in a few cases are separations achieved by exploiting differences in the kinetics of transfer) and that the distribution coefficient defines the resultant of a whole gamut of relevant equilibrium constants, then by identifying the specific equilibrium which is controlling in a case or type of case, one should be able to build up a family of extractants which would provide continuity in properties and therefore provide for freedom of choice.

Metal extraction

The studies around recovery of metals have probably filled more pages than any other topic in LLX; indeed, the success in PD for metal recovery in a limited number of cases has also probably been more pronounced than for any other example.

It is instructive to examine why the popularity and why the success. First of all, as ore quality goes down so it becomes necessary to resort to hydrometallurgy, and hence to the recovery of the metal values from dilute aqueous solutions. Liquid–liquid extraction presents great versatility because the presence of two phases implies separation even though the mechanism of transfer from one phase to the other may differ from case to case. Depending on the selected system, this can be utilized to promote separation of desired from undesired materials; furthermore, the energy of transfer is generally low which is important when feed streams are dilute.

Distribution by solvation is largely concentration controlled, hence this would not be a good choice as mode of recovery from dilute feed streams. Specific chelating or complexing systems are more likely to be favored in such cases. This has generally been the case in regard to recovery of metal ions from hydrometallurgical leach liquors.

Academic studies of interaction between metal ions and organic compounds have been pursued intensively; for years this was a fertile field for analytical procedures; hence a considerable bank of information exists from which to draw models of suitable reagents. Great efforts have gone into reagent synthesis and modification.

Liquid–liquid extraction in general has benefited from the metal extraction success story. Thus mechanisms, kinetics and equilibria have all been studied in this framework, and the findings are available for application in other cases of two-phase liquid separations. Furthermore equipment design, operation and control have all been boosted by metal extraction studies, and others stand to benefit from the lessons learned. Cation exchange concepts can be generalized for anions too, hence much is applicable in principle to recovery of acids and salts, although this has rarely been followed through.

The volume capacity required in a metal recovery plant may be vastly greater than is handled in other LLX processes, hence the operational aspects and design criteria serve as a way-out border for other studies. Negative aspects and problems too have been studied and therefore are of service to others in the field; thus problems of control, automation, simulation, solvent quality, solvent loss, interactions with subsequent process steps, all these can be generalized and fitted into other contexts.

The success story has had negative aspects as well, since R & D funds have usually had strings attached, hence the tendency has been towards intensive in place of extensive study, so that original approaches to systems and processes outside the framework have been limited.

Two different periods can be seen distinctly in the story of metal extraction. Early work was based on the ion exchange reaction using essentially water-insoluble acids as extractants. The basic ion exchange equation is:

$$\overline{HR} + MX \rightleftharpoons \overline{MR} + HX$$

The equilibrium constants for combinations of various M's with various R's will be pH controlled, hence pH of the aqueous phase at equilibrium controls the system and permits separations among cations. Since reversibility is purely pH controlled, it was necessary to neutralize the H^+ liberated to the aqueous phase for extraction to proceed. This step had distinct disadvantages, and led therefore to the second period where complexing agents were sought with formation constants high enough over a workable pH range to obviate the need for neutralization.

When metal ion separation and recovery utilized carboxylic acids as cation exchangers with pH control as the mode of separation, stripping required a competing acid or a competing base.

The order of the cations is essentially fixed for particular extracting acids. The difference between extracting acids relates to the strength of the acids, and this controls the pH scale for the cation separation, but not the order of separation—the whole cation order moves up or down the pH scale as a function of strength of the extracting acids used.

Subsequently, since copper recovery seemed to be the item of maximum interest, specific reagents were sought for copper. This program has been a triumph of the inter-disciplinary efforts of process development, organic reagent synthesis, marketing and mineral processing. Unfortunately, these speciality reagents, or specific reagents developed for copper extraction, do not enable extension of the procedure to other metal recoveries. Thus LIX® reagents when applied to other ions will behave as any cation exchanger, maintaining the fixed normal order of extraction.

For copper extraction the LIX® reagents represent a family of graded properties which suit the range of conditions encountered in copper processing. On the other hand, the successful extractants used in uranium recovery appear to be unconnected one with the other. The reason for this difference, even in the case of these two metals which are firmly established and accepted in the liquid–liquid extraction/metal recovery area, can be postulated to derive from the fact that it was the reagent manufacturers who opened up the extraction field for copper, while the uranium producers or users have essentially developed the uranium extraction field. What we seek now is for researchers to open up the extractant field for other metals, by relating to some graded order of relevant properties. This brings one to the relative basicity of the metal ions which is a natural scale, the need then being to define a counterpart scale of extractants.

Such an attempt has been made by using the amine salts as extractants on the one hand, or by setting up a family of acid extractants covering a range of pK's, or by utilizing the interaction of extractant and modifier to satisfy specific equilibrium constants requirements. The use of amine salts for cation extraction is well known, but these have not been regarded as generalized systems. Actually the scale of amines and amine–diluents covers a wide span, which should make this a system of considerable versatility.

Cation exchangers do indeed permit the setting up of a scale for reagent selection as a function of pK_a's of the reagent acids, the order being sulfonic acids/phosphonic/phosphoric/carboxylic; this expresses itself as a pH scale of selectivity.

It is clear that extraction by ion exchange implies the reverse in the stripping step where the ion is released from solvent phase back to an aqueous phase. The driving force for transfer will need to be different in the stripping and the extraction; usually it is the H^+ level that is the driving force for the stripping reaction.

$$\overline{MR} + HX \rightleftharpoons MX + \overline{HR}$$

Another means of changing the acid strength of the reagent between extraction and stripping so as to cause the reaction to go to the right is by changing the diluent; in other words, by utilizing reagent/diluent interaction.

The study of amine systems as extractants for acids seems to have been covered in an organized manner, i.e. the characteristics of the amines have been defined as a class. When one considers that amines are the largest class of organic bases, it makes sense that this has been so. This indeed means that one can fairly well select an amine according to what one wants from it—base strength, solubility, etc.

With acids as extractants this seems to be much less so, possibly because the acid strength of a mono-basic carboxylic acid is only slightly dependent on the C-chain length; also, above a limited number of C's, the solubility in water is not that much influenced by chain length. For a family of acids of graded strength, one cannot therefore really look to carboxylic acids; instead, the family must derive from substituted carboxylic acids or from esters of multivalent acids or from C–? bonds, e.g. C–P, C–S. The interesting fact is that even in resin cation exchangers only C–C and C–S bonds seem to have been exploited (i.e. there are strong cation exchangers—sulfonates, and weak cation exchangers—carboxylic acids) so there is not even a good analogy which can be followed by going from solid to liquid exchangers.

III. The LLX Separation Process — Technological Feasibility

Verification of the LLX Separation Process

Generally, a process operating in a liquid flow regime will be operated in the continuous mode, usually in steady state. Steady state verification of a flowsheet covers a whole range of aspects, "chemical" and "technological". Steady state can be defined on various levels which are not necessarily mutually inclusive. Thus steady state may be related to flows, with constant total in-flow, out-flow, and holdup, and/or to compositions.

In liquid–liquid extraction, without considering mechanisms of transfer, we can assume that a flowsheet will encompass at least two steps—the "extraction" step and the "stripping" or "release" step. Although the "stripping" or "release" step need not necessarily be a step of liquid–liquid transfer, let us for our purpose at this time assume it to be so.

The material balance for the system will, therefore, cover both steps in sequence, at steady state. As long as one is referring to the transfer back and forth of one component only, steady state is a simple concept. However, as soon as more than one component transfer, independent of or dependent on each other (or one another), steady state may not be attained simultaneously for all the transferring components, and non-steady state in one component may influence the transfer of another, obviating overall steady state being attained. If there is considerable difference in the stripping of a second component compared to the first, the second component may accumulate in the extracting liquid phase. This is a very important aspect of flowsheet verification, namely the build-up of components in one of the liquid phases which, for example, may affect the physico-chemical properties of this phase thus interfering with steady state operation. The maintenance of "solvent" quality is therefore an important aspect of flowsheet verification.

A flowsheet can be developed by an organized testing program aimed at defining limits and identifying constraints. However, once the flowsheet has been delineated, there is the need to verify it. Presumably the transfer and separation of the major components will have been used as the basis for the flowsheet, so one can assume that a part of the data is well based. However, the question of minor components, their distribution and behavior may be more difficult to predict directly from a limited test program. The reason for this is that the distribution of minor components may be strongly interactive with the concentration and distribution of the major components. Thus in a system in which distribution of major components are strongly concentration dependent, this can lead to back and forth transfer of minor components, with peak formation before steady state is reached with respect to these minor components.

The problem of build-up in the solvent of extracted minor components due to their high distribution compared to the major components may impair the solvent quality, e.g. the selectivity and/or the capacity of the solvent, in the course of a continuous cyclic operation.

Build-up in a cyclic operating system is the result of interacting effects on distribution coefficients or an inordinately high distribution coefficient of a minor component compared to that of others in the system. The former type of build-up depends on *back and forth transfer* without exit, the latter is a case of *one-way traffic*. General descriptions of the two cases can be represented without designating specific components. Very few such cases have been described in the literature, mainly because they are hard to detect without an operating model. A collection of equilibrium data alone will rarely show up such a case unless one is well aware of the basic aspects and therefore able to devise suitable experiments for showing up and locating the build-up effects or at least the likelihood of such build-up. Rarely have these effects been exploited, although unexpected benefits can in fact be derived by such exploitation, and unusual problems can be avoided when there is awareness and proper design.

Suppose, as shown in Fig. 30(a), we have a component A which distributes by virtue of the presence of B, and that $K_{DA} > K_{DB}$, but that $[B] > [A]$; suppose A is the desired product. The distribution of A will be much higher in the presence of considerable B, hence in order to obtain A essentially free of B one needs three steps, extraction, back wash to remove the co-extracted B, and wash or strip to recover A and liberate the solvent for recycle. The quantity of solvent used in E will be adjusted to the distribution coefficient of A so as to extract essentially all of A, but only some fraction of B. In the back wash the quantity of aqueous phase used will be determined by the level of B in the extract and K_{DB}. In order not to deplete the extract while passing through back wash, the aqueous feed will be drawn from the next step which is the wash for A recovery. This aqueous feed contains A which will in part transfer to the solvent phase as if back wash were an extraction step for A, but a stripping or washing step for B. The crux of the matter is that whereas in extraction, the extraction of A is influenced by the presence of the major component B, in wash this is no longer the case, hence the concentration of

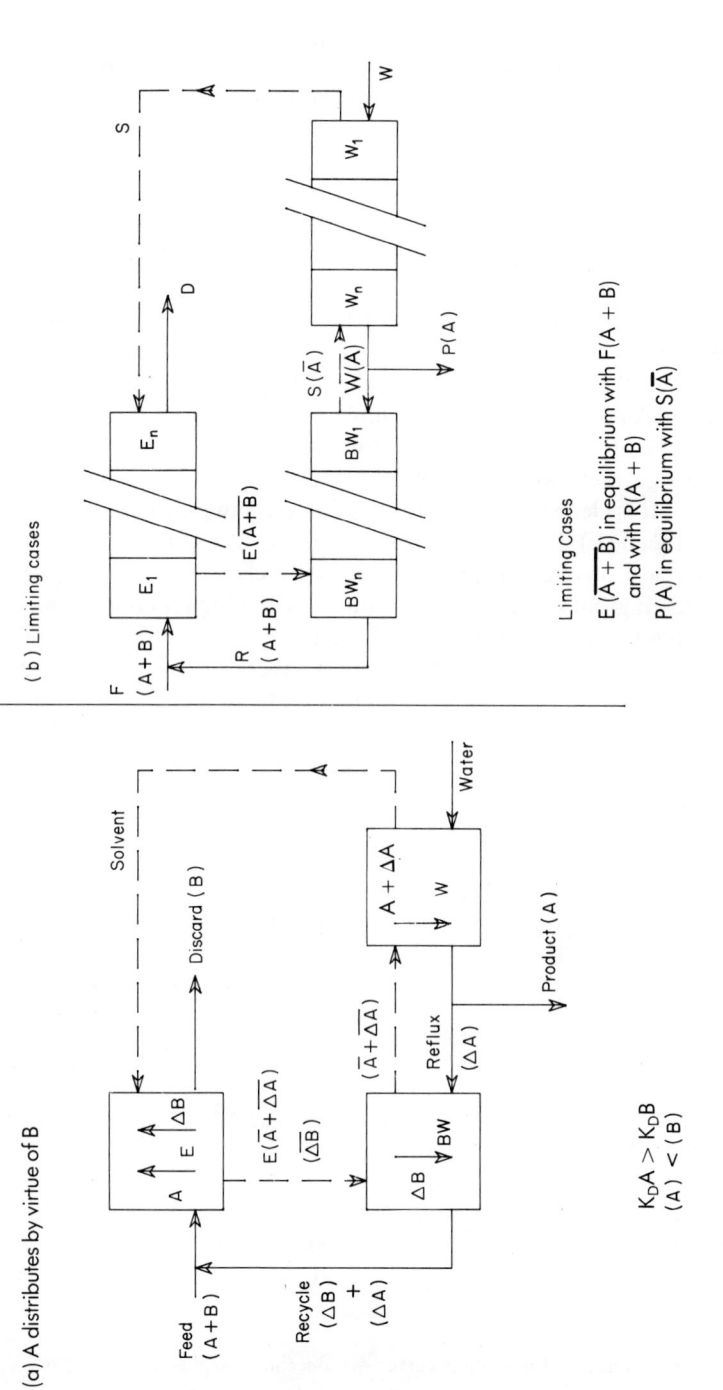

Fig. 30 Build-up by back and forth transfer (a) A distributes by virtue of B; (b) Limiting cases

A in the exiting aqueous phase from wash will be much higher than in the aqueous feed to extraction; this then explains the back extraction of A into the solvent passing through the back wash. A, therefore, builds up by the back and forth transfer of A between the wash and back wash steps, eventually attaining some limiting concentration. This is perhaps more clearly presented by looking at the limiting equilibria controlling the various steps, as shown in Fig. 30(b).

In the limiting case $E(\overline{A + B})$ is in equilibrium with $F(A + B)$; similarly $E(\overline{A + B})$ is in equilibrium with $R(A + B)$ hence $R(A + B)$ approaches $F(A + B)$ in composition.

At the other end, in the limiting case $W(A)$ is in equilibrium with $S(\overline{A})$, hence $S(\overline{A})$ is in fact the controlling concentration in the system. In a case where the presence and level of B in feed influences the transfer of A from feed to extract, we can exemplify the situation graphically as shown in Fig. 31. It is clear that the concentration of A in product (P) will be greater than in feed (F), if B is the factor influencing transfer.

The case of one-way traffic is similarly clear. Take a case of M which extracts in great preference to N, but N is at a higher concentration than M in the feed. The volume of solvent used in extraction for full extraction of N will therefore greatly exceed the volume necessary for free extraction of M; hence unless steps are taken to remove M from the system it will remain in cycle in the solvent, building up until it attains the equilibrium level in E

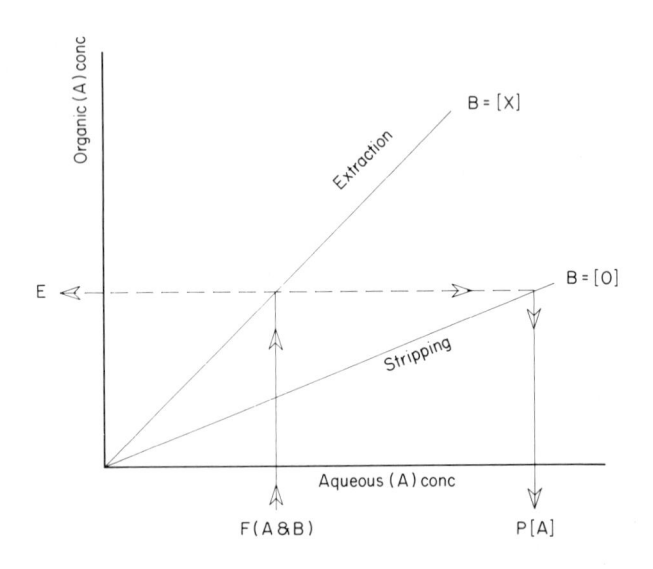

Fig. 31 Equilibrium curves—B concentration affects A extraction

when it will come out in the wash, but only after considerable accumulation at this point; the F and W would have the same composition in both components. This "one-way" traffic can be avoided by a pre-extraction step, which utilizes the quantity of solvent required for full extraction of M.

Technological verification is a serious aspect of any R & D project—not less so in LLX. This verification, required for decision making in regard to implementation, relates to various aspects, process, operation, equipment selection. The best technological verification is an industrial-scale operating plant, but clearly this cannot be a valid approach to technological verification of a new process.

Restricting ourselves to LLX processes, let us consider possible problems and the aspects which are likely to need verification.

The LLX flowscheme for separations in a complex multi-component system will usually be interwoven and integrated. Inconsistencies and disturbances which may be noticeable but considered irrelevant on the scale of R & D work may take on considerable importance in plant operation. Thus precipitates or crud which collect at the interface may seem insignificant in a separatory funnel or in a bench scale mixer settler, but may be disastrous on full scale. A real, though low, vapor pressure of solvent components in no way interferes with laboratory or bench scale test, but may have considerable economic or ecological bearing on an operating process. When a composite solvent is used differences in mutual solubility of components and water can lead to solvent disproportionation resulting in a composition drift.

A minor component which itself seems harmless even though it bleeds through to product may undergo slow secondary reaction giving rise to derivatives which can prove to be serious contaminants.

Feed solutions which have a natural variability, being, for example, of leaching, mining, fermentation origin, may cause problems if the minor components vary more than anticipated, or if components begin to appear without being anticipated. The more experienced the process developer, the less likely that significant signs will be brushed aside or ignored. Nevertheless it is necessary to decide what type, scale and duration of tests are required in order to give an acceptable level of technological verification of various aspects of a new process.

Technological verification for operation will be directed mainly to control aspects, and to the degree of swing which can be permitted.

Scale-up considerations are serious. The scale of testing is important for verification, from all points of view. On what scale should the process be checked out; on what scale should the operation be reviewed; what is the scale for equipment testing?

Since the scales required for verification of various aspects may differ widely, it is clear that an integrated plant model in the form of an operating pilot plant can be useful only if it is at least of the minimum scale for the most demanding aspects being considered. Such a pilot plant will be a costly undertaking and yet may not give answers to all the relevant process and operating questions. An alternative may be to test each aspect separately, on the scale it requires, and not as an integrated whole. If production of a representative quantity of product is required, it may be essential to run a model pilot plant to be sure that the quality produced is representative. In this case, operational and technological verification can perhaps be integrated.

The simplest approach to verification is by checking distribution coefficients in a cross current operation, in conjunction with a limiting conditions test. Cross current can be run in two directions and each direction places accent on a different aspect of the system.

On the basis of these equilibrations a scheme for the purpose in hand can be drawn up. Let us assume we have arrived at an outline of the process separations that will be required, such as pre-extraction, extraction, stripping, or whatever. We can check each step in various ways. If we have the limiting conditions test, and the cross current which gives the range of distribution coefficients for the operation across the concentration range required, the first thing to do is to see whether a graphical (or mathematical) analysis will indicate the number of stages required for some desired degree of separation or recovery. In a multi-component system a stage-to-stage calculation would normally have to be made if there are interactions between solutes; alternatively, simplifying assumptions can be made with the aim to enable a two-dimensional graphical procedure to be used for the analysis. Having determined the number of stages required for a particular extraction or recovery, the next step is to do a simulation of a countercurrent operation, using the number of stages postulated, and paying attention to the separation of phases, so that the stages will be theoretical stages as far as possible.

Once the extraction has been shown to follow the expected design, one can go on to the next operation, in a similar manner. The difficulty, however, in many cases is that one must generate the solution in one operation for use in the next one, and this becomes very laborious in a laboratory simulation.

The alternative is a bench scale continuous unit which will generate the intermediate solutions continuously. All that need be demanded of such a bench scale continuous unit is that it be reliable without artifacts, so that results are real. If, in addition, the bench scale unit shows up certain aspects which were not brought to the fore in the multi-stage system, this is all to the good. Thus precipitates, crud, encrustations, emulsions, foams, may absolutely not be observed in separatory funnels, but may appear in a continuous unit after some time. The time must be related to the holdup and the feed rate so one can get an idea of the significance of the observation. However, a bench unit is not aimed at this aspect, so one should realize that many phenomena may not show up at this stage but may come to light much later.

It is important to remember that real observations or phenomena have real significance and should not be passed over or brushed aside, however slight they may seem to be. What is insignificant on small scale may become very problematic on plant scale.

After the bench scale testing, one can assume that the basic flowsheet will have been verified. At this point decisions must be taken as regards the next stage of verification required. There are aspects relating to the solvent (e.g. stability and quality) which need to be verified. There are long-term aspects like cumulative effects which cannot be seen or sufficiently appreciated in laboratory simulation or in short bench scale tests. Two paths are open: one is to set up a pilot plant and operate it for a long enough period to be sure that cumulative effects have at least come to light, even if they cannot be expected to show up in their full magnitude; the other is to postulate effects and to devise a definitive limiting test to check the point under consideration. Thus very careful solvent stability testing can be done, anticipating likely instability paths, but these will not necessarily cover all the possibilities or eventualities which may derive from a certain juxtaposition of factors. Even a large pilot plant, representative of a full scale design, may mislead, because it will be a distorted model in regard to surge or cycle time, or to time at a particular temperature or acidity, or to penetration of air, or to loss by volatilization, etc.

Mass transfer and coalescence are aspects which need study, but which cannot in fact be separated from the type of equipment which will be used in a particular plant. Thus mass transfer rates can be determined, for example as a function of agitation which is a function of the agitator/vessel combination. However, a completely different type of contactor may show different mass transfer rates; similarly, coalescence characteristics can be measured but the use of the measurements will depend upon some correlation with equipment characteristics. Equipment selection is strongly connected with a knowledge of the flexibility and the sensitivity of the process. There are a number of ways in which equipment selection and scale-up can be handled, and these will have a direct bearing upon the R & D program for process development.

Process Development

Introductory

Liquid–liquid extraction implies transfer of solute from one liquid phase across a boundary to another liquid phase. This transfer has two aspects; one is "basic", relating to the mode of transfer, the other is "applied", relating to the means whereby the transfer is achieved. In the simplest context the former can be defined as chemistry, the latter as chemical engineering. No solvent extraction separation or transfer can be developed to the stage of technical feasibility, let alone process implementation, without the active inter-play of these primary disciplines.

In liquid–liquid extraction the separation is "system" controlled, hence a variety of systems can probably be conceived even for the separation of a specific component. Such systems will be as similar to one another, or as dissimilar, as one may wish. Not all, however, will be equally desirable nor equally feasible, so an acceptable choice must be positively made, if implementation is eventually to be attained.

Process development for LLX processes

Liquid–liquid extraction has been applied in a variety of fields for all sorts of purposes. The decision, in principle, to incorporate process development in a specific research framework is a policy decision at management level. Once a decision has been reached to go ahead with PD, subsequent decisions will be mainly technological or combinations of technology and management. Success will depend on close contact and understanding between the two, and on appropriate planning and organization.

Irrespective of the field or the technology, the development of chemical processes must go through a succession of stages. The individual stages may vary in significance according to the process involved; the order of the stages is not rigid but all must be covered in some way, if the relevant information is to be obtained which together will constitute the process. The aims of process development must be firstly to reach the point where a well based economic evaluation can be made; secondly to define the degree of uncertainty, as far as this is possible *a priori*, so as to enable management decision to be reached in regard to implementation; thirdly to furnish the required guidelines for basic engineering and design, and lastly to provide liaison between development and implementation.

Since it is our thesis that PD for LLX processes is no different in essence from other PD cases, the relevant stages can be listed in a general way for a comprehensive PD program. This has been done in Table 16 assuming a process of average complexity. In practice, LLX processes may show varying degrees of complexity according to the system and the type of separation desired. The complexity usually derives from the number of transfers that are required in order to attain the separation, and from the nature of recycle streams within the overall system. The number of theoretical transfer stages within a transfer operation does not contribute to process complexity, unless there are a multiplicity of streams entering and exiting. Operating complexity is not necessarily the same as process complexity, so operating control problems need to be viewed separately from the PD proper.

There may not be a consensus as to the delineation of the development stages of the PD program and the aim of each stage, but this is relatively secondary; the important aspect for consensus is that as one proceeds from stage to stage, one should be laying the building blocks for the flowscheme and eventually for the process as a whole.

Although, as shown in Table 16, process identification falls essentially outside the defined area of PD, it still is worthwhile to spend a little time considering the aims and constraints of process identification. The separation desired needs to be specified since this will help to describe the process being aimed at. Even in an extreme case where applicability of new reagents or solvent types is the motivation, evaluation is always according to what can be achieved. The separation desired and the liquid–liquid transfer need not be identical. Thus in a two-solute system in which A is the value to be

TABLE 16. Typical stages in a liquid–liquid extraction process development program.

Process identification		Specification of separation desired Identification of transfer required Selection of mode of transfer Choice of solvent type Test of chemical feasibility
Process development	Process research Process study Bench or pilot study Process package Liaison	— Specification of limits — Definition of flowsheet — Format and parameters — Parameters for technological design — Specifications of process — Guidelines for basic engineering — Contact during basic engineering and process implementation

and B is the accompanying valueless component, it may be equally valid to transfer A away from B, as B away from A.

The selection of the mode of transfer is of considerable significance for PD. Extraction systems have been classified as follows:

a. Distribution of simple molecules.
b. Extraction by compound formation.
c. Extraction by solvation.
d. Ion-pair formation.

Once one knows what is to be transferred one can select modes of transfer which are possible, and then the one which seems preferable. Choice of solvent type follows directly from this leading then to the test of chemical feasibility. In some cases chemical feasibility may be demonstrated in a most simple manner. Let us exemplify this by a metathetic reaction:

$$MA + HB \rightleftharpoons MB + HA.$$

If, from a stoichiometric mixture of MA and HB, the solvent extracts H^+ preferentially over M^+, and extracts some A^-, then irrespective of how much B^- is extracted, the reaction has in fact been demonstrated as going from left to right.

When PD starts one assumes that one has a selected system to work with, and that this system does what one has set out to do, but the limiting framework has not been defined. Specification of limits is the first stage of process development, here called process research. According to the nature of the process, the accent will be placed on different aspects; generally, however, during this stage one should derive sufficient distribution data to be able to define limits of extraction. One should have made comparisons within the family of solvents or reagents to see what aspects are favored by which members, so as to make a reasonable, more specific selection. Also one should be able to calculate separation factors, hence one can define the transfer steps involved, thus going over naturally from the process research to the process study stage.

The process study is probably the heart of process development, conceptually certainly, even though not as regards effort and expenditure. The aim of the process study is to define the process by laying out the flowscheme, fixing format and working parameters. The more complex the scheme needed for eventually attaining the desired separation, the more the process study will depend on the necessary input from those with background in relevant disciplines. This is also the point at which originality in manipulating solvent extraction concepts can be displayed, hence it is the most challenging. Since this is so, it is important that decisions should not be taken in ignorance of constraints imposed by other disciplines. This means also that some level of process costing must be initiated as early as possible. Almost as soon as the basic flowscheme can be visualized and the primary material balance prepared, an order of magnitude economic evaluation should be made, so as to identify areas for study and development.

Let us consider a process study for the metathetic reaction referred to above; let us assume that the process research has shown that the highest extraction of A^- will take place when the aqueous phase is essentially saturated with MA and when H^+ is high, which must imply B^- being high too. This finding naturally leads to the need for understanding the phase diagrams

$$/M^+ /A^- /H^+ /B^- /H_2O /$$

and

$$/M^+ /A^- /H^+ /B^- /H_2O / solvent/$$

as a function, say, of temperature.

In this process study, the phase diagram together with distribution coefficients and separation factors, are the basis upon which the flowscheme will be built.

As long as we stay within the framework of separation by LLX there is only a limited number of possible steps. Their selection and arrangement will be based essentially on relative concentrations and separation factors, on solvent modification possibilities within the framework of solvent compounding, and on the extent to which temperature and/or pressure are transfer variables. Heat effects must be carefully tested. Small scale testing cannot show up such effects. Calorimetric measurements must accordingly be made and primary heat balances calculated.

The wider the scope of the chemical input in all its aspects, the more likely that the flowscheme will be simple, elegant, original and useful.

The aim of the bench and/or pilot study is to check out the selected flowscheme, as far as possible under continuous operation, and to define the parameters for technological design and for a well based economic evaluation.

The scale on which this study is to be performed is open, the choice depending on many factors, and relating to many aspects. There are aspects which can be studied equally well on laboratory scale by using multi-stage countercurrent simulation, but there are questions that can be answered only in true continuous operation. In Table 17 typical aspects have been listed for laboratory simulation and for continuous operation. This, however, still does not answer the question of scale. One of the problems of bench scale operation is the difficulty of keeping steady state in small units. Also there is the danger of being misled due to the incidental character of operation in the unit selected for the test. Rarely will one expect a small unit to be a scaled down version of the full industrial plant. Even the equipment

TABLE 17. Achievement in laboratory simulation versus continuous operation.

Multi-stage laboratory simulation
— Full stage efficiency
— Profile of conjugate phase compositions
— Basic compositions for steady state operation
— Primary material balance
— Good indication of trends and preferences
— Physical observations, but no estimate of physical problems

Continuous operation
— Importance of procedures for filling and starting
— Possibility of working with recycles
— Check of material balance
— Possibility of seeing accumulations
— Possibility of adjusting dispersion type for good separation
— Primary entrainment aspects
— Checking effects of feed variability on operation
— A basis for defining process and operating control

type in the small unit need not represent the final selection; at best a small continuous unit will indicate the direction to be followed in making such selection.

The true purpose of the bench or pilot study is to define the parameters for the technological design so as to be able to specify the process. In Table 18 a summary list is presented of what this stage is expected to contribute to the process development, and what it will not provide.

TABLE 18. Contribution of piloting to process development.

Positive
— Verification of flowscheme
— Confirmation of material balance
— Security as regards unexpected operating phenomena, but not necessarily so for all eventualities
— Good indications for plant operation, process control, process sensitivity etc.
— A good model for starting up a plant, first filling, shutting down etc.
— Information necessary for preparing guidelines for basic engineering of the process

Negative
— No data for heat balance; calorimetric measurements must be made specifically
— Usually no design data for equipment, these must be acquired in a separate study, with or without vendor participation
— In the normal scope of the study, the continuous small scale unit will not provide sufficient product for marketing studies

In liquid–liquid extraction processes there are essentially three basic equipment types, namely mixer settlers, columns and centrifugals. Every solvent extraction operation entails liquid–liquid contacting and liquid–liquid separation. The three types of equipment differ from one another as regards these two crucial steps and each may also be evaluated differently according to whether one is looking at the dispersion or the coalescence characteristics of the system, and which one is overriding. Furthermore there may be a strong interaction of the process on the equipment selection and of the liquid system on equipment behavior.

Equipment and Process

The considerations which influence the choice of equipment for a process entering implementation for the first time will very likely be different from those which will obtain later when the process has been established and its weaknesses or strengths are more soundly defined.

Without relating to order of importance, aspects to be considered are:

Costs—investment
      —maintenance/down-time costs
Scale-up
Turn-down
Mass transfer, rates, stages
Phase separation
Crud handling
Entrainment and its dangers
Profile/distributions across stages
Start-up
Shut-down
Control and reliability
Ease of operating a mix of equipment types
Inventory
Expansion

There are various ways of approaching equipment evaluation and selection. One is to be aware of the basic characteristics of the different equipment types, and of how they operate as a function of system properties, then to see where the system under consideration fits in. The second approach is to try to identify the overriding consideration(s) in an anticipated process, and to see which equipment types would favor such considerations. It is of course clear that "overriding" considerations, too, will change for all sorts of reasons, hence equipment selection today may be invalidated tomorrow. This means also that the first implementation of a process should not necessarily dictate the equipment selection for the next implementation although often that is exactly what happens. It would seem to be mandatory that after the first implementation, all the aspects listed above should be reviewed once more and "overriding" considerations once more identified. Equipment is then to be reviewed in the light of the latest developments and fitted to the current "overriding" considerations.

There may be problems which derive from accumulations in continuous plant operation, which could not be anticipated *a priori*, but will be known only when they have been observed to be so; furthermore such conditions cannot readily be simulated in a short term experiment or test. It is clear that the operating plant does furnish the conditions, since it is there that such accumulations are being observed. Any test program for equipment evaluation must therefore attempt to relate to these and other known plant phenomena or conditions. This brings us to the question of the validity of pilot testing for systems which have, for example, accumulation tendencies. Short term tests may not show up these accumulations nor the effects, if any, of such accumulations. This could pose a problem in equipment evaluation but need not pose a problem once the fact is known, since undoubtedly a valid experiment can be designed to take into account what is known.

Basically one can say that the more one knows about a system the easier it will be to design valid experiments to answer valid questions; the less one knows about the system the more one tends to put one's trust in a larger scale test, i.e. in some scale of pilot unit.

Where process and equipment are interactive the scale-up of the equipment may not retain the "similitude" of the interaction, particularly when scale-up know-how is not "in house". Testing a variation of a process in a "pilot" unit not suitable for, or not scaled for, a particular step, may lead to erroneous conclusions or may delay the conclusions by virtue of irrelevant problems related to the equipment and not to the test.

A summary approach to matching process demands to equipment characteristics is presented in Table 19.

TABLE 19. Process interactions with equipment selection.

Required I—Matching equipment characteristics to process demands

Process
— Basic concept and format:
  Convenient combination of flows to achieve specified separations by mass transfer
  Interactions, recycles, phase ratios, concentrations, flow directions
  Efficiency of each step for its stated purpose
  Physico-chemical properties of the LL system, the range of change between separating pairs

Table 19 *contd.*

- Yardsticks:
    Degree of separation
    Product quality
    Minimum losses

Equipment choice
- Aim:
    Mass transfer and phase separation
- Yardsticks:
    Efficiency in separation
    Capacity
    Yield
- Constraints:
    Physico-chemical system
    Interactions
    Investment
    Capital charges
    Variable costs
    Utilities

Required II—Identification of physico-chemical constraints in relation to equipment type
    Evaluation of types of equipment, their constraints and potential

Technology
- Phase contact, for mass transfer
- Phase separation
- Significant physico-chemical properties
    Specific gravity
    Viscosity
    Interfacial tension

Rates
- Mass transfer, controlled by
    Transfer mode, concentrations
    Viscosity
    Interfacial area
- Interfacial area, controlled by
    Contact film area
    e.g. dispersion into droplets
- Dispersion controlled by
    Power input, physical properties
- Separation controlled by
    e.g. droplet coalescence
- Coalescence controlled by
    Dispersion type O/W, W/O
- Dispersion type controlled by
    Phase ratio
    Interfacial tension
    Mode of energy input

Testing, Scale, Validity, Piloting, etc.

The more one knows about systems of the type one is considering, the less likely is one to need extended testing on a larger scale. However, each time when even an experienced person starts on a new process there will be new angles and unanticipated problems. It is not that one does not know how to set about designing experiments, it is that one does not know all the questions that need to be answered, hence one cannot design the tests for answering unasked questions.

Process validity has many facets, and all of these must fit together for a valid whole. How, then, can one hope to have an acceptable process picture on which to base major decisions in regard to implementation and design, *before* one has an operating plant? But, one must have such a picture or there will be no justification for a plant in the first place.

One way to handle the problem is to consider any single feasible scheme as being sufficient in the first place and to check it through as far as possible, while keeping options open for change as new aspects come to the fore. The scheme that is selected as being feasible must be self-consistent and self-contained, including all recycle and bleed streams. Here one comes to the first problem, since recycles may cause build-up to some steady state level which may be difficult to anticipate *a priori*. It is clear that a computer program which can be used for reiterations until steady state is reached can handle the problem easily. However, in order to have a valid program it may be necessary to do so much generation of reference data that this is simply not justified for a single feasible flowsheet. The story is different, of course, for *the* flowsheet, i.e. the one that has been proven and selected as the basis for a plant; certainly, then, a program aimed at easily estimating steady state as a function of process changes is fully justified. For the initial flowscheme, however, it may be simpler to use the concept of "limiting conditions" so as to estimate the limits of levels which are possible and thus build a "limiting" material balance.

Naturally the next step must be checking out this limiting material balance, on laboratory scale, bench scale, or pilot, whichever is convenient. The problem always is the time required for attaining steady state in minor components—the best test is a material balance—"what comes in must go out" once steady state has been reached. There are cases where this requires weeks, months, etc. depending on total volume in the system. In a test system holdup can be kept to a minimum by restricting volumes, thus steady state can be attained more quickly; however, this makes the system much more sensitive to changes and fluctuations, so that in fact limiting steady state concentrations may not be reached.

There have been cases where unexpected problems have come to light only after the process has been implemented on plant scale. At such times it is common to say that piloting was insufficient. However, this may by no means be so, since only a fully representative scheme will replicate the full plant, and rarely is a pilot fully representative. Two things are very important: one is to be sure of the validity of the basic scheme, so that no unanticipated changes will invalidate the whole approach; and the other is to be ready for unexpected problems with all the tools to tackle and solve them.

Rarely will the new problem relate to the main component or the main separation. It is the minor aspects which are far more difficult to study *a priori*, and these, therefore, take on dramatic proportions initially in the first plant.

Here we come to another aspect, and that is the unwillingness at management level to change, once a basic flowsheet has been implemented. There is a tendency to follow an established pattern, and this may limit the possibility of preventing recurrence of the problems encountered in the first plant. One of the reasons is that PD essentially stops when implementation or design takes over, so that subsequently problems are "solved" but not "obviated" or "eliminated" because the "trouble shooting" approach is used instead of the PD approach. Basic development should not stop at the stage of implementation nor should liaison between PD, design and implementation come to a halt.

The problem of scale of testing prior to design is somewhat different when it relates to equipment instead of process. The interaction between equipment and process implementation may be very great. Often the test work on some standard piece of equipment furnished by the vendor will be the basis for the vendor's scaling-up, without the tester knowing exactly the scale-up factors. Subsequently, this becomes a reference, which in fact it is not, so later review of equipment is done against this first selection, instead of against the process itself.

When equipment is being tested for transfer by the primary mode it is important, though not necessarily obvious, to see what the implications are for other modes of transfer, e.g. carry-over or entrainment. Different types of equipment may show up differently when this is the way in which behavior is evaluated.

Raw materials or feed streams used in PD test work must be representative and "fresh". It is far better to draw feed from a production line than to store a quantity so that all test work can be done with the same starting material.

Equipment and Process Interactions

In choosing equipment for a LLX operation it is fully accepted that the properties of the system should have a bearing on the choice of equipment. Thus interfacial tension, viscosity and difference in specific gravities are such limiting properties. However, the interaction of process and equipment is not so commonly accepted.

First of all, even relating to the properties listed above, these may change dramatically through the process, i.e. by virtue of the transfer or separation itself. In other words, characteristics of feed streams may not be determining, but rather the exiting streams may have significance. These changes are process related, and they have a real, direct effect upon equipment selection, capacity and scale-up.

However, process–equipment interaction may be much more far reaching, particularly in complex multi-component systems with a multiple of transfer steps.

Dispersion type controls mass transfer rates and coalescence rates, but process controls the direction of mass transfer. These aspects may be contradictory, and not necessarily favorable for a particular choice of equipment.

Equipment of one sort may be very satisfactory for a particular process step, but not for another. This means compromise, or using a variety of equipment types, each as best choice for a specific step. A mix of equipment will require careful matching of the process requirements to the equipment characteristics so as to ensure smooth operation.

Some aspects that should be taken into account when considering a mix of equipment types are listed in Table 20.

Let us consider some aspects. In a particular process step entrainment say of solvent in the aqueous phase may be absolutely unacceptable; however, if this constraint is the only unfavorable aspect of an otherwise favorable choice of a specific equipment type, it may be worthwhile to handle the entrainment aspect separately from the process step itself. This, of course, is not always favorable, hence one may have to forgo the optimum choice in order to cope with the specific problem.

TABLE 20. LLX equipment mix.

Considerations:
  Inter-operational surge
  Response to process variability
  Off-specification streams
    Intermediate and recycle
    Product, raffinate
  Planned maintenance shut-down
  Turn-down capacity
  Manpower
    Supervision
    Operation
    Control

In a different example it may take considerable time to arrive at steady state with regard to some minor component in multi-stage countercurrent operation—hence, any piece of equipment which has an appreciable holdup but cannot maintain a concentration profile on stopping may be a bad choice, whereas the same problem of lost profile may be acceptable if holdup volumes are very small.

In a system in which there is considerable water transfer, specific gravity differences may become small at some point in a countercurrent process so that conditions for phase separation may be different from entry or exit conditions.

Equipment on-line time, maintenance, cleaning, etc. may be strongly process dependent. In other words equipment type that is absolutely suitable from capacity point of view may not hold its own because of down time resulting from characteristics of the process. Thus, for example, if we have a system which becomes saturated with respect to A, by virtue of extraction of B, A will precipitate at some intermediate point in the process. Some types of equipment may allow the solids to proceed along with the liquid; other types may do the reverse, holding back the solids and thus causing accumulation, which in turn requires shut-down and cleaning. Without cleaning, capacity may be greatly impaired.

In some cases equipment modification may overcome the problem, in others this equipment type may simply be ruled out.

Testing equipment for selection and sizing will differ considerably, according to whether an operating plant is on hand from which to draw process streams. Furthermore testing of a model unit, said by the vendor to be of sufficient size for scaling up, may nevertheless not give reliable results if aspects of accumulation and its effect on operation cannot be taken into account. Similarly variability in process streams cannot be taken into account unless there is an operating plant from which to draw such variable streams, or unless there is a real understanding of the cause or source of variability, and the likelihood that this will not be an operating variable. For example, build-up of impurities in the solvent phase which affects coalescence rates, or changes in solvent composition due to differences in solubility or volatility of components, or insufficient stripping, requiring a change in phase ratio, all these may have direct effect on capacity. If these are not taken into account, under-design may be the result.

Coalescence and Entrainment

In mass transfer between two phases the contact area across which the transfer can take place is one of the controlling parameters for mass transfer rates and hence for efficiency of utilization of equipment. Phase separation after mass transfer is no less important. In LLX these requirements usually express themselves in efficiency of dispersion and coalescence, which are strongly interactive concepts; poor coalescence and separation in a multi-stage countercurrent system will tend to cause a decrease in efficiency of the transfer; this explains, for example, the studies devoted to axial back flow in column extractors and to coalescence rates, etc. However, the accent is usually on efficiency of transfer of the component at which the extraction is aimed; this applies to the use of enhancers for drop to drop coalescence, the utilization of the effects of wetting characteristics, centrifugal separators, etc. All these aspects are macro problems. However, the micro problems which result from entrainment, but which do not truly show up in connection with macro mass transfer studies, can easily be overlooked.

In a countercurrent system, entrainment at the two ends is important, not because of effects within that countercurrent system itself, but because of what comes next. For simplicity let us consider first of all entrainment generated only in those last contacts where the two phases respectively leave the mass transfer countercurrent system. The implications of entrainment may be entirely different in the two cases, and need to be examined and considered separately.

In a coalescence test it is well known that the dispersed phase will give a clear coalesced phase, while the continuous phase may remain cloudy, due to a very fine dispersion which cannot coalesce under the conditions obtaining for this operation. On the basis of these facts, in a system of discrete stages (e.g. a mixer settler system), it will be recommended that the correct dispersion type be promoted so that the clear coalesced dispersed phase leaves while the cloudy continuous phase proceeds along the multi-stage system. This phase going along the system will eventually be the phase exiting at the other end. Hence the question to be asked is whether the fine entrainment which generates the cloudiness will be coalesced and removed as this phase moves through the system, either as the continuous phase or as the dispersed phase. Quantitatively there is little significance to such entrainment but qualitatively it may be of considerable importance, for the following reasons: at the extract end this fine dispersion will, in the limiting case, be dispersed feed phase, while at the raffinate end, in the limiting case, it will be the lean or depleted solvent phase that is retained as the fine dispersion. In a system where the purpose of the extraction is purification, i.e. where a material is to be separated from an impure environment, fine entrainment of impure feed can have an extremely negative effect on final product quality. At the raffinate end, even when there is little economic importance to the low quantity of solvent actually comprising this fine entrainment, it may be important ecologically, depending on the fate of the raffinate.

Various approaches can be used for attempting to eliminate the problems of fine entrainment. Thus high speed centrifugation can be considered, or a dissolution step in the form of a wash or scrub step, or a diluent wash, or adsorption by passage through an active material, or promotion of additional coalescence with the aid of wettable surfaces; in some cases a temperature change may cause phase separation and this separating phase will act as scavenger; in certain cases, particularly for aqueous raffinate, a steam strip may remove dissolved as well as finely entrained solvent by vaporization, provided volatility is favorable. Of course, steam stripping cannot remove non-volatiles dissolved in the entrained solvent, or composing the solvent phase entrained. Certain of these operations can be incorporated in the liquid–liquid extraction system, others are better considered as part of the next downstream operation, particularly where such a step is required in any case. Thus an adsorption step may be required for color control, so if the entrainment removal can be incorporated in the same adsorption system this may solve the problems. It should be clear that the quantitative levels requiring removal may be in the "parts per million" range, which will usually be orders of magnitude lower than the measurements made for the bulk transfer, in estimating extraction efficiency. What is being discussed is not the "back flow" or "axial flow" which is the concept that decreases the stage efficiency in LLX equipment, but possibly more an entrainment related to the physico-chemical properties of the system (at the time of writing this can probably not be said definitively). Thus, interfacial tension or the presence of trace quantities of surface active material may be important in some cases.

Certainly this micro aspect of entrainment separation will normally not be part of an equipment study program, even when phase separation is under consideration. Thus the technique used for determining entrainment will usually depend on a laboratory centrifugation test, while fine dispersion often can be determined only by micro analysis for some compound selected as the trace indicator.

This aspect of LLX may be of importance, too, when selecting a technology for a specific operation. Thus when comparing solid and liquid ion exchangers, it is clear that the liquid passing through a resin bed will not be contaminated by a stable, suitably washed resin; this applies both to raffinate and to stripped product. On the other hand, contamination of product by raffinate can indeed take place, since washing of resin even in the best case may not be efficient enough to avoid some micro retention and therefore micro contamination. Similarly, when precipitation with subsequent solid/liquid separation is used, this too may not prevent micro contamination of filtrate with precipitate, but quantitatively this may be much lower because of the relative volumes involved. This has odd consequences—in some cases, even though LLX may be a highly selective technique, there may still be low level contamination, due not to distribution, but to fine entrainment.

Demixing/Second- or Third-Liquid Phase Formation — Coalescence and Entrainment

In biotechnology the use of biphase aqueous systems for enzyme separation and recovery from fermentation media has been receiving attention. These multiple phases can be generated in various ways, usually directly from the fermentaion broth by adding phase demixing components with or without changing the temperature. The question then is to what extent is coalescence and entrainment controllable by the operation conditions imposed for the demixing?

In LLX, dispersion of one phase in the other is the accepted procedure for generating the interfacial area required for mass transfer at an acceptable rate. At a fixed phase ratio the nature of the mixing and the characteristics of the two-liquid system will determine which phase is dispersed; in order to invert the dispersion it will be necessary to change the phase ratio or the mixing regime. The drop size distribution in the dispersed phase will be a function of the properties of the system, the phase ratio, and the power input according to the type of mixing used. Coalescence will be a function of the dispersion which has been generated, and of the system characteristics.

In biotechnology, whenever it is possible to generate the conjugate phase so as to operate in a closed steady state cycle, as is usual in other LLX operations where mass transfer occurs across the liquid–liquid interface, then the accepted concepts of two phase contacting and phase separation will apply.

However, where components are added or parameters changed to cause demixing, so that the second phase is generated from the first phase, the normal variables of LLX contacting and phase separation do not seem necessarily to apply as such. An analogy needs probably to be drawn with other cases where supersaturation causes nuclei of a phase to separate, such as occurs in boiling or crystallization. These nuclei need then to grow to droplets of acceptable size for coalescence to start—this would point to the necessity for "conditioning" before phase separation. On the other hand, demixing and solute distribution are certainly conceptually separate, especially at low solute concentrations. Thus demixing could occur by adding the necessary components to water, even in the absence of fermentation products which are to be separated by virtue of this demixing and generation of a second phase. It should be clear, however, that nucleation and droplet growth may be influenced even by low concentration trace materials present in a fermentation broth, especially if these have any surface active characteristics. As the concentration of products of the fermentation increases, these need to be considered and regarded as additional components which may have an effect on generation of an additional phase. When the fermentation product is a macromolecule and a range of molecular weights are present it may be difficult to specify the number of components in relation to number of phases and degrees of freedom. If the presence of the fermentation product affects the viscosity considerably, this will have a definitive effect on droplet formation and growth, and coalescence and phase separation. The mode of distribution or partition between phases needs to be defined here, too, as in any other LLX separations.

Clean-up in a LLX System

Solvent quality, product quality, accumulations, etc. in a LLX system need to be addressed in the light of the system and of the process; these two need not necessarily coincide. Thus solvents can be classified as selective and/or blocking, as desired, or as specific; furthermore the same solvent may be selective in one case and specific in another.

In some instances solvents may be regarded as transfer agents, in other words the transfer is as follows:

1. $AqI \xrightarrow{A} Solvent \xrightarrow{A} AqII$

The same solvent may be selective for another component in the same context, but not a transfer agent for this component. This can be represented as follows:

2. $AqI \xrightarrow{A+B} Solvent \underset{B}{\rightarrow} \xrightarrow{A} AqII$

If the solvent is a transfer agent also for B, we shall end up as follows:

3. $AqI \xrightarrow{A+B} Solvent \xrightarrow{A+B} AqII$

In the case of Example 2 there is typically accumulation of B in solvent, while in 1 and 3 this will probably not be so. However, in the case of 3 the product will be contaminated by some quantity of B which has passed via the transfer medium/solvent.

Clean-up of solvent may have different levels of implication in the three cases, as also will product clean-up.

There are cases where the solvent will block the transfer—thus if the solvent is specific under the transfer conditions, there may be no danger of transfer—thus crown ethers that specifically transfer K^+ by size will not transfer Mg^{2+} or Na^+. Similarly in Cu^{2+} extraction by LIX® oxime reagents, Ni^{2+} will not transfer.

Similary a cation exchanger will exclude the transfer of anions, i.e. it is *specific* for cations, but it will only be *selective* within the category of cation exchange itself.

Solvent clean-up takes on an important aspect in all LLX systems, but not necessarily for the same reasons in different cases. Thus two different reasons could, for example, be product quality in one case versus solvent quality in another, or solvent quality as it relates to the extraction characteristics, i.e. the mass transfer in a sense, or to phase separation characteristics. In the extraction of uranium it became clear that some other element was being co-extracted but not released, so that capacity was steadily decreasing. This entailed introducing a clean-up step which released the co-extracted element from the solvent at some time, so as to maintain solvent capacity. In the case of Example 3 it may well be that the K_D's for A and B are very different, so if B can be kept low enough relative to A in the solvent, it will be transferred to AqII at so low a level that this will not represent a quality problem. In this case a "back wash" step on extract may be the answer.

$$4. \quad \text{Aq} \xrightarrow{A \gg B} \text{Solvent} \xrightarrow[\text{AqIII}]{d(A) + B} \xrightarrow{B} \text{AqII}$$

Contamination of solvent so that its coalesence characteristics are affected is an entirely different problem. The contaminating components may be transferring by an entirely different mechanism, and clean-up therefore will need to be different. Thus trace quantities of a surface active agent, the source of which may lie far upstream, can have a disastrous effect on phase separation. Clean-up may require a distillation of solvent, or a washing step under particular conditions, etc. Slow reaction with solvent or in solvent, generating a component which influences phase separation may have an insidious effect on phase separation–coalescence characteristics.

Make up of solvent components to compensate for physical loss can serve also as a clean-up step in the sense of maintaining some level of impurity. Make up of solvent components due to physico-chemical losses, e.g. solubility in outgoing aqueous phases, volatilization due to azeotrope formation, complex formation leading to reduced solubility in solvent phase, and so on, do not contribute to clean-up unless the impurities accompany the components themselves.

The problem caused by dissolution or distribution of components of the feed which distribute by an entirely different mechanism from that of the main component for which the system is designed can affect the solvent quality and/or product quality. A good example is the organic matter of apatitic phosphate rock which will distribute to an organic solvent and cause various types of problems subsequently. A second example is proteinaceous or carbohydrate materials in a biotechnological feed to extraction.

Solids in LLX Systems

The appearance of "crud" or interface solids in facilities for recovery of uranium or other metal values such as copper has bugged the technology all along. Generally it is assumed that crud is due to very fine particles deriving from the dissolution step and being carried along in the aqueous feed, until caused to coagulate or flocculate by virtue of wetting by or adsorption of solvent phase. Generally, therefore, attention has been directed to clarification of the aqueous phase prior to the LLX step, or to removal of crud from the interface in the phase separation equipment.

Actually crud formation may result also from solubility changes caused by reactions on contact with the solvent phase or by the extraction itself, so it cannot be prevented simply by having a clarified feed solution. In certain contexts these very changes of solubility by virtue of reactions or by extraction, may be the purpose of the LLX itself. This leads to the necessity to understand precipitation or crystallization by reaction and puts crud formation into a somewhat different perspective.

The behavior of crud may be very different in different types of equipment according to whether the equipment moves the crud on or retains it, whether it promotes flocculation or breaks up the floc, whether there is time for crystallization or instead there is rapid fine precipitation.

Process Control

Even in the simplest two-step LLX process, there is a certain logic to selecting the control points. This simplest countercurrent, multi-stage LLX system can be presented as shown in Fig. 32, assuming a solvating solvent, transfer being wholly concentration controlled. Let us assume a case of HX being extracted from an impure aqueous feed of medium concentration; let us also assume that the impurities do not extract into the solvent selected, at least not beyond the limit which can be accepted in the product. Also let us assume that the presence of these impurities does not affect the distribution of HX in any way. (It should be stressed that almost all of these assumptions can only be relatively true, but sufficient for our case of demonstration.)

The aim of the extraction will be to recover the maximum quantity of HX from the feed, hence a primary control point will be the level of HX in the raffinate. The ultimate level of HX in the raffinate is controlled, however, by the level of HX in the lean solvent, leaving the stripping, hence an additional primary control point will be the level of HX in the lean solvent.

The second aim of extraction will probably be to recover as concentrated an HX product as possible in the heavy product leaving stripping. This then

Fig. 32 Two-step LLX process

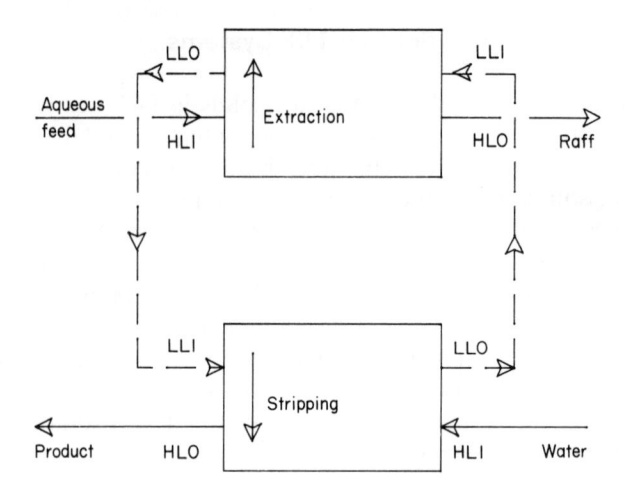

LLI = Light liquid in
LLO = Light liquid out
HLI = Heavy liquid in
HLO = Heavy liquid out

too will be a primary control point. However, ultimately, the concentration of HX in the product out of stripping must be absolutely limited by the concentration of HX in the loaded solvent entering stripping, which is in fact the loaded solvent leaving extraction. Here again the ultimate control of the level of HX in loaded solvent must be the level of HX in the primary aqueous feed to extraction.

The picture we get, therefore, is that, equilibrium-wise, in this system everything is interrelated, so in fact we are no nearer real process control.

Let us now reverse the analysis. Provided the equilibrium curve is available for the distribution of HX between the aqueous phase and the chosen solvent over the whole range of concentration from zero up to the feed concentration, it is possible to determine the interrelation of concentrations of "in" and "out" streams, for extraction and stripping as a function of phase ratios and number of theoretical transfer stages. This can be seen from the graph in Fig. 33 where a fairly normal equilibrium curve for HX extraction by a solvating solvent such as alcohol is shown. Such an equilibrium curve is best prepared by cross current contacting of the aqueous feed with the selected solvent, analysing conjugate phases for HX at each stage of contact. This is preferable to using compounded solutions made of pure HX

III. The LLX Separation Process—Technological Feasibility

Fig. 33 Extraction and stripping with common equilibrium curve

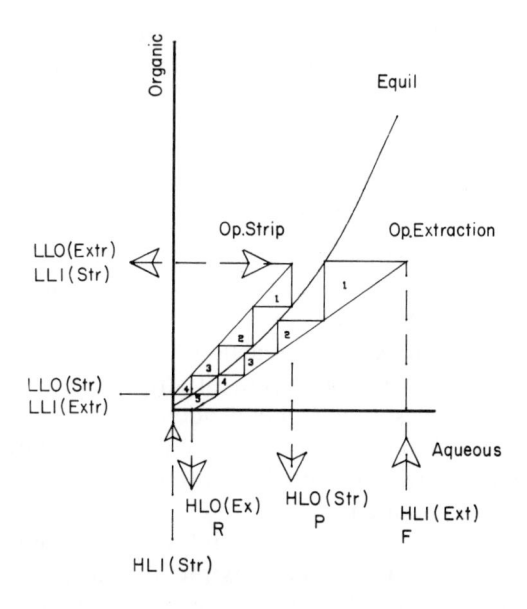

Extraction = 5 Stages 75% recovery
Stripping = 4 Stages 60% concentration
max (75% concentration)

and water, because of possible non-extractables in the feed which may influence distribution.

It becomes clear that the shape of the equilibrium curve is controlling, with full extraction more demanding than full stripping. Decisions relevant to design, which then become fixed points for operating control, such as feed volume and the phase ratio, expressed as solvent feed volume, must first be taken before process control can be logically constructed. Also a cost analysis must be made to balance raffinate losses, against investment and against product concentration. Only in this way can one arrive at a well based approach to standard operating procedures and also to guidelines for process control, and corrections for process drift.

The reason for this strongly interactive situation is the fact that in the assumed case, the equilibrium line is assumed fixed, essentially independent of parameter changes within the anticipated operating range. When this is not the case, the situation may be appreciably different.

Let us now take a case of cation exchange, using an acidic feed and a cation exchanger with a stability constant in the stronger acid region, so no neutralization in extraction is required. Stripping is accomplished by reverse ion exchange using an acid. Let us use $M^{2+}X^{2-}$ as the example, with a suitable $\overline{HR}$ reagent. Then we have the following:

$$M^{2+}X^{2-} + \overline{2HR} \rightarrow \text{Extraction} \rightarrow \overline{MR_2} + H_2X$$
$$\leftarrow \text{Stripping} \leftarrow$$

The M^{2+} recovery will be a function of the number of stages provided, as well as the quantity of reagent fed per unit of M^{2+}, at the conditions specified for the particular case, within the permissible range of acidity change in extraction due to the ion exchange reaction proceeding to the right. However, the concentration of product out of stripping due to the reaction proceeding to the left is essentially independent of the concentration of the MX in the primary aqueous feed, being wholly controlled by the H_2O/H_2X ratio in the acid fed to stripping, and by the free acidity out of stripping and the number of stages provided.

In this case, therefore, process control in extraction can be based on the residual M^{2+} level in the raffinate out of extraction, although this may not be such a critical control point if the acidic raffinate is returned to leaching. The stripping will be controlled by the concentration of M^{2+} in product, and by the level of unstripped M^{2+} in the recycle solvent. However, the strong driving force for extraction deriving from the concentration of the free acidic reagent in the returned solvent phase in a sense decreases the importance of its residual M^{2+} level.

In a flowsheet in which a scrub step is included, i.e. where the loaded solvent is to be scrubbed with water or with a bleed of product, the control becomes more specific. Thus, if the scrub is to remove a co-extracted minor component of lower distribution than M but which strips by the same mode as the M, then control aspects apply similar to those in the main stripping. However, volume of acidic bleed will be very important here in order to keep to a minimum the quantity of recycle M accompanying the impurity out of the scrub. Since M in extract is, as it were, in unlimited supply, it is clear that concentration of M in the aqueous phase out of scrub will always be at a maximum, hence volume controls the material balance in this operation.

When the scrub is aimed at removing an impurity which extracts by a different mode from cation exchange, e.g. when an organic impurity is present that extracts by partition, the control of the scrub will need to be the residual level of the impurity remaining in the solvent phase leaving the scrub step. On the other hand, when scrub is aimed at removing entrained feed, the control may relate to any non-extracted component in the primary feed.

There is a different aspect of process control which must be considered and that is how to correct drift in operation. For this to be valid, it is necessary to do an analytical review of interactions in the various steps which can result in drift, arranging them in a logical order. Thus if "a" is occurring, this may first of all be because of "b", but if "b" is in order, the reason may be "c", if "c" is not the cause, the reason must be "d" and so on. It should be borne in mind that in some cases an incorrect assessment of the cause of drift, and therefore a non-valid correction, may cause the situation to deteriorate still further and may cause drift in other parts of the process as well.

Process control is ideal for playing "games of tactics". While operating control and process control are distinctly separate conceptually, they are in a sense like cause and effect, and therefore interactive. The "games" entail a "what if" or "suppose that" approach. This is probably the ideal place to introduce a computer once a process has been outlined in its salient aspects. Previously it was necessary to "play the tactics game" intuitively, inductively and deductively, using step changes to simplify the analysis. Now the rapid reiterations of a computer can make this a much more practicable approach. This is especially the case as regards drift and the possible effects of incorrect interpretations on subsequent process steps.

In LLX systems, on-line analytical control is not common. However, as on-line procedures are developed for other systems, no doubt these will have application in LLX systems too. Here the problem is the presence of the two phases in contact and the necessity to have full phase separation for valid analysis. Probably the one case where on-line analysis could be introduced with a minimum of effort is where centrifugal separators are used since phase separation is at its best and holdup is the lowest of any of the equipment types presently in use.

Inventory of Surge in LLX Systems

A LLX flowsheet will usually be based on flow control. This requires that some minimum volumes be maintained at strategic locations within the flowscheme to take care of normal variations and intended changes. The question of solvent losses and solvent make-up is a very relevant one in LLX systems, both for process technology and for process economics. This accordingly requires regular summation of solvent inventory in the plant. In processes where volume change throughout the LLX is small, this is less of a problem than in cases where considerable volume change takes place in loading and stripping. One of the most straightforward approaches to this problem is to use the lean solvent, in surge, as a measure of inventory, provided that steady state is being maintained in the system, and that interface locations are fixed. The seriousness of these constraints in relation to the accuracy of the inventory is very much connected also with the equipment type being used for mass transfer, since volume holdup can vary enormously from type to type. Where a single component constitutes the primary solvent, the question of composition drift or change is not relevant to the inventory. However, where a compounded solvent of several components is used, the steady state composition in the lean solvent surge becomes very important.

The lean solvent surge tank may have a second very important task to fulfil, and that is as a check on solvent quality, and for solvent make-up. Most processes will have a solvent clean-up step incorporated into the flowscheme. This permits a directional flow as regards solvent quality if this is significant, process-wise. Thus, for example, if solvent make-up is periodic, there may be no benefit in not feeding to the solvent surge tank, especially in the case of compounded solvents where make-up quantities of separate components may not be in the same proportion as the basic ratio in primary solvent feed, since process losses of components may differ according to their separate properties. The important point in this type of situation is to keep the composition of the bulk surge constant. In the case of a single-component primary solvent, it may well be advisable to have a constant feed of make-up solvent, and to feed this into the system at a point where benefit will derive from the quality of the primary solvent. This may, in fact, apply equally well in the case of a compounded solvent as far as the positive clean-up procedure is concerned. Here, too, it may be worthwhile to draw solvent for clean-up from the bulk surge, but not to return the cleaned stream back to the average in bulk surge, but to feed it at the point where solvent quality is most significant.

Various procedures can be applied to check on solvent quality. Here there are two aspects to be considered. One is the steady state maintenance of quality of the bulk solvent surge, both as regards overall composition and as it relates to solvent behavior at significant process points. The other is an estimate of the efficacy of the clean-up step, in an absolute sense, i.e. "in" and "out" of clean-up, and relative to the accepted quality of the bulk surge. If clean-up cannot maintain the desired steady state quality of the bulk surge, this may mean that a material balance on impurities is not being maintained, hence a larger bleed or drag of lean solvent to clean-up is required, or that the clean-up procedure does not take care of the overall impurity types being accumulated or generated in the system.

Most LLX processes can tolerate a certain solvent make-up per unit of production, from the overall economic point of view, and this will have been considered in the review of the basic viability of the process. No LLX process can tolerate a steady deterioration of solvent quality, whether this expresses itself as a steady decline in the capacity of the solvent for the material being extracted, or as a decline in the quality of product exiting from the LLX operation, or as a deterioration in the capacity of the mass transfer equipment, whether this relates to the kinetics of mass transfer or to the kinetics of phase separation.

In continuous, steady state operation, it is important to have a reliable point for control. This is particularly important in systems with a variable primary aqueous feed stream, since normally only limited variability can be tolerated in the product stream, notwithstanding the feed variability. This means that the solvent must be capable of maintaining the quality specified irrespective of average excursions in feed. This applies, too, to feed concentrations; the solvent must be able to accommodate reasonable concentration changes, at the production rates specified and in the available mass transfer equipment. Normally the solvent surge will take care of these variations within design limits.

Evaluation of Alternatives. Optimization vs Timing, Safety, Economics

It is difficult to generalize approaches for evaluating alternatives, because practical evaluation cannot be considered out of context. However, one can take a number of typical situations, and compare the modes of evaluation in each case. One must consider inter- and intra-evaluation, e.g. by comparing a LLX possibility with alternative non-LLX options, or by comparing two LLX processes. Let us start with two LLX processes, both of which satisfy the basic requirements of the separation. Aspects for comparison are listed in Table 21.

A review of all the above aspects may or may not show up points which make one or the other solvent preferable. Non-process aspects therefore come up for consideration; for example, the questions of optimization and timing, how much experience is there relative to either solvent? Are there reasons to consider specific equipment in one case or the other, if so, are there delivery problems which have bearing on the choice? What are the safety aspects, are they as stringent with both solvents? Solvent losses have a bearing on the process, its technology and its economics. Starting from the obvious, the economics (the cost of the solvent and its separate components) will reflect on the process economics. Its physico-chemical properties will impose constraints which must be translated technologically; these also are important in the process definition. In addition, aspects of ecology and pollution in relation to solvent losses may impose stringent process and technological constraints.

In general, no solvent process can be economic unless the solvent is in

TABLE 21. LXX—Aspects for comparison.

Definition of separation required, between solutes A and B:
  Some minimum recovery yield required
  Some minimum product quality required
Now take solvent systems X and Y:
  Type of solvent, e.g., BPt, azeotrope, stability, viscosity, interfacial tension, gravity
  Protic ⎫ Acidic⎫ Molec. Wt.
  Aprotic⎭ Basic ⎭ Diluent (?)
  Mutual miscibility with water as function of temperature
  Distribution coefficient of A as function of concentration and temperature
  Distribution coefficient of B as function of concentration and temperature
  Separation factor of A and B
  Phase ratios, by material balance
  Number of stages required in extraction for X; . . . for Y
  Number of stages required in stripping for X; . . . for Y
  Necessity to scrub before stripping
  Solvent removal from raffinate
  Solvent removal from product
  Water balance
  Extraction of minor components, solvent clean-up
  Mass transfer rates
  Coalescence rates for W/O or O/W
  Dispersion type of natural phase ratio
  Need for recycle to change dispersion type
  Volume changes in extraction
  Volume changes in stripping
  Is either solvent already in use in the facility?
  Solvent availability and cost
  Need or desirability to use diluent
  Rate of solvent make-up

---

closed cycle. In limited cases solvent can be separated from the extract by distillation, but usually it is preferable to separate the solute from the extract by a liquid–liquid transfer of solute from solvent into an aqueous phase. Solvent losses relate then to residual solvent in the two aqueous phases of the LLX process. The level of loss that can be tolerated in the raffinate may be economically or ecologically controlled. The mode of recovering the solvent may rely upon evaporation, secondary extraction, and/or adsorption. Evaporation with vapor recompression is an economic procedure in many cases. Recovery of process solvent from an aqueous process stream requires the utilization of a second solvent of considerably lower solubility in the aqueous phase than the primary solvent. Adsorption on carbon or on speciality adsorptive resins may be used effectively. The more dilute the initial feed, the larger the volume of raffinate per unit of product, hence the more this operation takes on importance even for low cost solvents.

In the case of aqueous product, solvent elimination may relate more to product quality than to the economics of solvent losses; here again distillation, extraction, and/or adsorption may be considered. Thus in certain cases an additional extraction step, e.g. with an essentially water immiscible low boiling point solvent could be used; again the evaluation will be largely economic.

Solvent losses must be viewed in two ways—chemical losses, due to side reactions, as a function of process conditions or operating conditions, and physical losses due to entrainment, poor phase separation, spillage, etc. In any system it is necessary to study the stability of the solvent and its components in relation to possible routes for reaction, breakdown, etc. This will have a direct bearing first of all on solvent choice. Early in PD it is necessary to define extremes of process conditions so that solvent stability can be studied, at least at the limits.

Re-evaluation with Experience

It is not unusual for an integrated scheme to be worked out very carefully for a first plant, based on all the available know-how and experience. During start-up and running-in, problems begin to appear, some already expected in essence but not to the degree found in practice, some unexpected for various reasons. Problems which need urgent solution will be tackled "in plant context" and solved, while others which the plant can "live with", will be allowed to ride, even though in principle they appear to negate the basic concepts of the original system and its aims. If the second plant is to be erected shortly, it will follow the first very largely in concept, but with the primary problem corrected, while the other problems are handled within the design or in the standard operating procedures, but without the original concept being adjusted or adapted in any way. Now, in the second case, it may appear that the secondary problems are no longer secondary, since their significance begins to show up very clearly once attention is removed from the original primary problem. It is found now that the second plant cannot "live with" the problems by simple adjustment of operation. However, it is too late to make conceptual changes, hence *ad hoc* measures will be taken to enable the plant to "live with" the solution, while complying with requirements of quality, economics and environment, whichever was the reference for dissatisfaction. Assuming that such corrective measures do work, there is a good chance that the third plant will incorporate these measures, again without reviewing the basic concepts, so that it too will not be an integrated system based on the most up-to-date analysis, in the sense that the first plant was, at the time, being based then on the best that was known.

In doing a primary integrated LLX process design for plant implementation, a very careful mental exercise has to be followed in two senses, the first is within the scope of the LLX system itself, the second in relation to the overall process and the way in which the LLX system fits into the whole. The questions that were asked when approaching the first plant need to be asked and answered again, and in as basic a manner, for the third plant too. The exercise is represented in the following, by listing some questions that need to be asked and answered, or taken into account in the integrated design:

1. Why use LLX? Let us assume that this is a recovery/purification operation to displace a previous procedure. The LLX operation must give equal or better yields and equal or better quality at comparable points; it must also tie in equally well or better with the next downstream operations. It must permit a controllable steady state operation from all relevant points of view.
2. What is the main separation that needs to be performed and what is the choice solvent system for achieving this?
3. How are known impurities expected to behave with this chosen solvent system? Will they extract to any degree? If so, how will they be eliminated? What positive steps are to be taken to ensure that quality will be preserved, i.e. quality of product and quality of solvent?
4. What types of impurities are taken care of by the procedures proposed? What types will not be handled by these procedures? Is there a procedure that can prevent even unspecified impurities from slipping through? What solvent clean-up step will prevent accumulation of known type impurities in solvent?

It has been pointed out that there is only a limited number of operations which can be fitted together in an overall LLX system—similarly, there is only a limited number of modes of transfer, depending on the characteristics of the material being transferred and the characteristics of the two liquid phases. Similarly, also for solvent clean-up, there is only a limited number of operation types that can be performed, depending on the solvent characteristics and the nature of the impurities to be eliminated.

Taking the flowsheet as originally conceived, it should be possible to examine the steps that were incorporated then, to review the purpose and to compare this with what has actually been achieved, and then to relate this reality to the updated understanding of the system. In this way, it is possible to see whether the step in its original context is doing what the more recent understanding indicates needs to be done. If not, then it probably needs to be revised.

An alternative approach is to lay aside the present format and the initial approach, and to start as it were from scratch, but at the present point of progress and knowledge, i.e. once again to define the aims and the constraints, and to see how one would put the system together now, without prejudice, but keeping to the same solvent system so as not to lose the experience gained. If for any reason a different solvent system looks attractive, this should be regarded as a different exercise entirely.

There is one other point which needs careful evaluation: when is it desirable to mix operations instead of sticking as far as possible to one concept? Thus, for example, liquid ion exchange for major components may fit in well with a resin exchanger at some point, probably downstream, if specific removal of a low concentration minor component can be achieved.

Another aspect of considerable importance is to identify points at which drag streams should be withdrawn in order to maintain acceptable levels of accumulating minor impurities, how these streams should be tested, and where they are to return to process.

Optimization versus timing is an interesting topic for consideration. In a competitive society, timing is important. For a new technology in an established industry, if there is a real economic advantage, it is usually worthwhile to get into production as soon as possible; eventually competitors will catch up, either by developing something newer or better, or by finding their own route to the same achievement. The time advantage is therefore ephemeral and must be exploited while it lasts.

Optimization has two aspects, "*inter-*" and "*intra-*". Let us assume one has arrived at a *worthwhile* process and is proceeding with implementation. At this point *inter*-optimization in the broad sense is irrelevant. In other words, there is no point in seeking now for a *more worthwhile* process, since implementation is already being lined up. Eventually, certainly, after some period in production it will again be wholly acceptable to seek a more worthwhile approach.

On the other hand *intra*-optimization should be going on all the time, aimed at better yields, lower conversion costs, better quality, easier and more efficient control, waste reduction, less pollution, better utilization of equipment, improved capacity, etc. This *intra*-optimization is as valid in LLX as in any other PD work. Many improvement possibilities come to light once the plant is being operated. Then one can say, "maybe if we did this instead of that, we would obviate the other, which is causing problems". In the initial flowscheme design, this problem could perhaps not have been anticipated, hence a solution, even if trivial, was not sought. More far-reaching improvements, too, will come as the plant operates—thus, for example, some modification of solvent compounding and composition, a change in operating temperature in order to promote something, or to prevent something else. New ideas and better ideas usually come as a logical continuation of something else by recognizing the need.

In summary, therefore, during the stages of R & D and PD optimization is valid only to the extent that it reflects on economics, and to the degree that it can be introduced without in any way prejudicing the timing. If a major unexpected improvement comes to light during the later stages of PD or even during plant design, management will need to reflect and decide whether delay is justified in the specific case.

Safety in LLX processes is an interesting question. In an area in which organic solvents are not customarily in use, there may be hesitation and fear in regard to introducing such solvents, while in an industry where this is usual, the same step would not cause comment. Naturally, good design and plant practice must be part of the process concept, for one reason at least, that losses of solvent can probably not be tolerated purely for economic and ecological reasons. Still, safety codes are fairly well developed in most industrialized countries; if there is no direct code, then a suitable analogy should be sought. Again, if the use of solvents is entirely foreign to the field under consideration, the whole concept of adjusting the design in all its aspects, e.g. pumps, piping, sumps, etc. to this fact may meet with a negative response. On the other hand, aspects of safety related to the current technology will be applied without hesitation, even though the new process may in some cases make these obsolete.

LLX requires no different precautions from those to be taken in any process using organic liquids. This general approach will relate in particular to diluents and modifiers, where volatility, flashpoint, explosion point, etc. are normally to be considered. Active reagents may need to be looked at in a somewhat different way, as regards, for example, toxicity, allergy reactions etc. If there are any decomposition paths, these should be evaluated, and decomposition products regarded in the same way as the primary reagent. Air and water pollution are obvious aspects of importance.

Evaluation of a LLX process is no different from evaluation of any other process. The same difference in approach between primary processes and secondary processes will apply. There is a ground limit for a LLX process based on the volumes to be handled, which will exclude the use of the technology for a product value below some reference figure. Since equipment is modular in many cases, scale is not as significant as it might be in other technologies.

Comparison of LLX with Other Procedures for Achieving Similar Separation

Separation procedures in chemical engineering can be grouped as follows:

(1) distillation, extraction, absorption and ion exchange are essentially dependent on partition or distribution between two phases, as a function of concentration.

(2) precipitation and crystallization both depend on solubility limits and are therefore of a different family.

Since all of these operations lead to separation, it is legitimate to make comparisons—of similarities and of differences. In the final instance the choice will depend on efficiency and costs, as is always the case.

The concepts used for defining transfer systems in distillation, extraction and absorption are very similar, all depending on the concept of equilibrium stages. These lend themselves to steady state operation. The same principles can be applied also for ion exchange, and indeed this is done in the case of so-called "liquid–liquid ion exchange" but not usually where solid resin ion exchangers are used. The reason for this is mainly operational, since in the latter case unsteady state is usually applied, the resin being held in a fixed bed where kinetic, diffusion phenomena play a significant part in defining the operating system. It is legitimate, however, to regard multiple fixed beds as representing steady state, and this then makes resin ion exchange fit into the same "transfer stage" concept as the other three.

Not all cases lend themselves to a choice from among the four operations. Clearly, distillation implies reasonable volatility, extraction requires identification of a mode of transfer in some acceptable two-liquid phase system, while absorption is generally restricted to a gas/liquid system, and ion exchange, by definition, implies an exchange or an ion-pair reaction.

Assuming, however, that there is a free choice, say between extraction and distillation, or between liquid and solid ion exchangers, then fairly clear general guidelines can be given for making comparisons. Special aspects of the specific case may override these general comparisons, however. Comparisons will relate to fixed costs, variable costs, process control, operational control, efficiency, quality, losses, contamination. Secondary cost aspects may be plant area required, inventory of reagent required, speciality expertise available, similarity to familiar operations, etc.

In the uranium recovery area there has been a swing back and forth between LLX and Ion Exchange (IE). However, for copper recovery there seems little likelihood of reverting to IE since LLX has taken over. Possibly the difference in the two cases is the concentration in feed streams. There is much attraction in IE when levels are in ppm range, while the disadvantages of IE become clearer as the metal to be extracted goes to higher concentrations. Another difference between IE and LLX is the ease of continuous countercurrent operation in the latter case.

Energy-wise, the main transfer in LLX or IE is very low in energy consumption while distillation of course is inherently high, entailing a change of state. Subsidiary energy requirements may, however, far exceed the transfer energy, thus changing the picture entirely. Also, vapor or mechanical recompression may favor distillation. In IE, dilution is a serious factor; in LLX this may be so too, but there are expedients in LLX to overcome this, as for example when there is a favorable temperature effect on solubility of water in the solvent system.

Evaluation of LLX

The knowledge that "oil and water do not mix" is ancient; however, this is a long way from providing the basis for selecting LLX systems; in nature two liquid phases in mobile contact, with mass transfer across the liquid–liquid boundary, are at best rather rare.

It is, of course, not necessary to seek out naturally occurring LLX systems, since it is relatively easy to construct such systems by proper selection of combinations. The challenge, therefore, is two-fold—on the one hand to extend principles developed in regard to one specific system type to other cases of the same type, and on the other hand to define new system types with potential for application to a variety of cases, directly or by analogy and with modification.

The former challenge is more straightforward—the history of cation separation and recovery can be regarded as a good example. The second challenge is more difficult since it is not so obvious where to begin. In its simplest context LLX is a separation procedure, but this may imply the transitive operation "to separate" or the intransitive concept of "being separate". This brings us to two sets of questions, namely, "what is it that is to be separated, from what", and therefore "how", or "what needs to *be* separate from what" and what benefit can derive from this "being separate"?

In order to bring both approaches into a reasonable range for handling and practicality, it seems essential to define real cases which can then serve as generalized models for extension and expansion.

Generally, starting out with procedures for separation has not proven fruitful, unless desired separations are first identified, so as to check the appropriateness of applying the proposed procedures. On the other hand, after desired separations have been identified, it should be possible to define the ideal procedure for attaining this separation, or to compare various proposals for this type of separation.

When one regards LLX as a general tool, and when there is sufficient familiarity with LLX in a generalized context, it becomes natural to ask whether two liquid phases can serve the separation purpose in hand. Familiarity is a complex concept since it implies a number of aspects which may not customarily be regarded, or examined, in juxtaposition. How to favor the attainment of such familiarity is the problem that requires solving. At first glance, the simplest solution in all contexts is to regard LLX and its application as a group activity. Individuals, therefore, are required to have the general awareness but not necessarily the overall expertise.

In the case of other separation operations, it is not unusual to expect the equipment or reagent vendor to supply the expertise required. Thus if one wishes to separate by crystallization, one is justified in assuming that the vendor has the overall expertise, or if one plans an ion exchange operation, that the resin supplier will provide the overall expertise. In LLX this is not so, since equipment vendors or the suppliers of solvents and reagents will have only specific expertise. To establish applicability of LLX, first in principle, then by selection, and then by specific examination, it is necessary to know what *questions* are relevant, i.e. what one needs to ask, and then to seek the answers. This approach is the basis and the main justification for this book, and this is its integrated aim—to ask the correct questions, so that correct answers may be sought and found.

Recommended Reading

The number of books on liquid-liquid extraction published during the past two decades is small, although many papers have appeared in a variety of scientific journals. However, the best overview of developments in this area can be derived from the published *Proceedings of the International Solvent Extraction Conferences (ISEC)*; dates and venues of these conferences are listed here.

In writing this monograph on liquid–liquid extraction the intention has been to present an integrated approach to this powerful tool for separations from liquid media. Such an approach does not necessitate substantiating each individual statement by reference to published literature. In place of a multiplicity of separate, specific references, an up-to-date list of books on the subject has been provided here, as a guide to general reading in the area. Those requiring in-depth information on limited aspects will always have to make their own detailed literature reviews.

1. *Proceedings, International Solvent Extraction Conferences (ISEC)*
   1986 Munich, Germany
   1983 Denver, Colorado, USA
   1980 Liege, Belgium
   1977 Toronto, Canada
   1974 Lyons, France
   1971 The Hague, Holland

2. *Phase Equilibria in Chemical Engineering*
   Chapter 5. Phase Diagrams
   Chapter 7. Liquid–Liquid Equilibria
   Stanley M. Walas
   1985, Butterworth, London

3. *Partitioning in Aqueous Two-Phase Systems*
   Eds. H. Walter, D. E. Brooks and D. Fisher
   1985, Academic Press

4. *New Developments in Liquid–Liquid Extractors*
   Eds. R. B. Akel and C. J. King
   1984, AIChE Symposium Series No. 238, Vol. 80

5. *ACS Awards Symposium on Separation Science and Technology*
   St. Louis, MO. April 9–10 (1984).
   Ed. J. D. Navratil
   Separation Science and Technology **19** (11 & 12), 723–941
   1984, Marcel Dekker

6. *Handbook of Solvent Extraction*
   Eds. T. C. Lo, M. H. I. Baird and C. Hanson
   1983, John Wiley & Sons

7. *Solvent Extraction* Part I and Part II
   Eds. G. Ritcey and A. W. Ashbrook
   1979, Elsevier

8. *Design of Solvent Extraction Systems*
   R. Blumberg
   Separation and Purification Methods **8** (1), 45–71
   1979, Marcel Dekker

9. *Solvent Extraction Chemistry*
   T. Sekine and Y. Hasegawa
   1977, Marcel Dekker

10. *Recent Advances in Liquid–Liquid Extraction*
    Ed. C. Hanson
    1971, Pergamon Press

11. *Industrial Extraction of Phosphoric Acid*
    R. Blumberg
    Solvent Extractions Reviews **1** (1), 93–104
    1971, Marcel Dekker

12. *Partition of Cell Particles and Macromolecules*
    Distribution and fractionation—in aqueous polymer two-phase systems.
    Per-Ake Albertsson
    1971 (Revised 2nd Edn), Wiley-Interscience
    1986 (3rd Edn), Wiley N. Y.

13. *Ion Exchange and Solvent Extraction of Metal Complexes*
    Y. Marcus and A. S. Kertes
    1969, Wiley-Interscience

14. *Phosphoric Acid*
    Fertiliser Science & Technology Series, Vol. I, Part II

    Chapter 8.   Purification of Wet-Process Acid
                 Solvent Extraction: Chloride Addition and Temperature Cycling
                    A. Baniel and R. Blumberg
                 Solvent Extraction: Use of Partially Immiscible Solvent
                    A. V. Slack

    Chapter 11. Use of Acids other than Sulphuric Acid
                 Hydrochloric Acid
                    A. Baniel and R. Blumberg
                 Nitric Acid
                    R. J. Piepers
    1968, Marcel Dekker

Index

Acid extraction
  acid–ether two temperature solvating system, 103, 104
  acid–ether–alcohol, 104
  acid extraction, 139, 140
  equilibrium diagram, 100
  modifier–alcohol, 99
  modifier effect, 104
  operating time, 100
  solvating solvent, 101, 102
  tertiary amines, 99, 100, 102
  three-phase zone, 104
  two-acid separation, 102, 125

Bench scale
  equipment type, 116
  perturbations, 116
  process control, 116
  shut down, 116
  start up, 116
Biotechnology—LLX systems
  analogies, 50, 51
  bi-aqueous phase generation, 54
  classification possibilities, 58
  cycle steady state operation, 52
  demixing, 51, 166
  extract disproportionating, 54
  feed conditioning, 54
  fermentation, 55, 56
  high molecular weight materials, 49, 50
  low molecular weight materials, 49, 50
  reversibility, 51
  transfer phase, 50
  water rejection, 50, 51

Computer simulations
  computer aided process design (CAPD), 116, 127, 128, 129, 130
  control, 129
  optimization, 129
  process synthesis, 129, 130
  simulations, 126

Dilute solutions
  basic aspects, 27, 28
Distribution coefficients
  aided distribution, 4, 6
  partition coefficients, 4, 6

Equilibrium curves (distribution curves)
  acid/alcohol systems, 110, 113
  amine systems, 110, 113
  extraction, 25, 110, 113
  implications, 33
  shape, 23, 24, 25
  stripping, 25, 110, 113
Equilibrium systems
  comparisons, 90
  equilibrium stages, 108
  Gibbs phase rule, 89
  McCabe–Thiele graphical procedures, 109
  multicomponent, 89
  multiphase, 89
  primary concepts, 90
Equipment
  attributes, 69
  characterization, 68
  choice, 66, 156, 157

Equipment — *continued*
  evaluation, 66, 156, 157
  positive tests, 68
  process demands, interactions, 69, 156, 157, 158, 161, 162
  scale-up, 149
  selection, 66, 68, 156, 157, 158
  testing, 160, 163
  types, 66, 156, 157
Evaluation
  economic, 67, 110, 175, 176
  LLX, 180, 181, 182, 183
  process, 110, 177, 178, 179
  safety, 175, 180
Examples, typical
  complexation, specific, selective, 136
  counterpart phase, 135, 137
  distribution, 133, 134, 135
  reversible transfer, 133, 134, 135
  separation factor, 133, 134, 135
  solvent efficacy, 135
  water transfer, 133, 134, 135
Extractants
  acidic, 141
  amine-salts, 106
  basic, 141
  chelating, 141
  ion-pair, 141
  major components, 106
  minor components, 106
  mixed, 141
  neutral, 141
  selective, 1, 2, 17, 143
  selectivity controllable, 4
  specific, 1, 2, 17, 143
  solvating, 141

Feasibility
  chemical, 78
  demixing, 78
  distillation analogy, 78
  technological, 145
  viewpoints, 78
Flowschemes
  advantages, 116
  disadvantages, 116
Flowsheet
  analogies, 114, 120, 125, 126
  auxiliary steps, 114
  build-up, 115
  continuous testing, 115
  costs, 28, 38
  counter-current simulation, 115
  delineation, 28, 81, 82, 84, 87
  internal cycling, 115
  main steps, 114
  material balance, 115
  partial vapor pressure analogy, 120
  purpose, 28
  rectification/stripping analogy, 12, 120, 125, 126
  recycle/reflux, 120, 125, 126
  targets, 28

Gibbs phase rule (in LLX)
  acid–ether/alcohol—3 phase zone, 104
  degrees of freedom, 41
  examples: acid–salt systems, 34
  invariant points, 34, 104
  multicomponent, 33, 38, 39
  multiphase, 33, 37, 38, 76, 89, 104
  phase changes, 75
  pseudo invariant, 37
  reacting, 104
  third liquid phase, 7, 76, 103

Interactions
  environment, 69
  peripheral operations, 69
Interactive solvent systems
  amine extractant, 105
  classification—chemical interactions, 128
  common principle, 105
  solute/solvent, 78, 79, 80
Interfacial tension, 22, 71
Invariant systems
  cross current, 113
  distribution coefficients, 113
  limited conditions, 113
  program of study, 119
Ion-pair extraction
  neutralization—stripping, 100
  heat balance, 100
  reversal of reaction, 113
  sensitivity to variable, 113
  systems: third phase, demixing, modified, 104

LLX operation
  analogies, 10, 14, 47, 48
  detriments, 73
  dilute solutions, 26
  economics, 66, 67
  engineering, 66, 67
  examples, 13
  extension, 46
  scope, 12
  separations, 2, 13, 46
  separation factors, 117
  solids, 169
  transfer characteristics, 2, 60, 66
  transformations, 44, 46
LLX process
  applications, new, 56, 57, 58, 59, 60
  chemical feasibility, 72
  constraints, 85, 86
  process development, 88
  process implementation, 88
  reversibility, 46, 136
  selectivity, 2
  significant parameters, 86
  specificity, 2
  strategy, 87
LLX purpose
  cation separation, 141, 142
  conversions, 104, 137
  metal extraction, 135, 136, 138, 141, 142
  reactions, 137, 138, 139
  recovery, 7, 104
  separations, 7, 13, 104
  separations between acids, 125
  separations between acids and salts, 126, 136
  transformations, 44, 45, 46, 137, 138, 139
  upgrading, 7
LLX systems
  basis
    diluent-free, 104
    water-free, 104
  classification, 81, 82, 128
  clean-up, 167
  comparisons, 90, 180, 181
  contamination, 168
  design, 84
  evaluation, 175, 182, 183
  format, 82
  ion-pair, 110

membrane, 21
modes of study, 89
optimization, re-evaluation, timing, 177, 178, 179
process study, 93, 99
product quality, 167
primary concepts, 90
solvating, 110
slow reaction, 168
solvent
  inventory, 175
  losses, 175, 176, 177
  make-up, 168, 175
  quality, 167, 175
  surge, 173, 174
LLX systems—interactions
  common ion effect, 140
  coupled transport, 21, 74, 75
  critical solution points, 46
  diluent–extractant, 36
  driving forces, 91, 92
  non-polar diluent addition, 140
  peripheral operations, 70
  phase changes, 75
  solute/reagent, 75
  solvent phases, 79
  solvent/solute, 79
  viewpoints, 78, 80
  water, 93
LLX transfer
  back extraction, 30, 31, 81
  driving forces, 91, 92
  extraction, 30
  mass transfer coefficients, 22
  mass transfer rates, 22
  pre-extraction, 30, 81, 125
  scrub, 30, 81
  solvent clean-up, 30
  stripping, 30, 81

Metal/cation extraction
  cation, separation, recovery, 135, 136, 141, 142, 143, 144
  specific reagents, 143
  selective reagents, 143

Operational aspects
  automation, 142, 143
  control, 142, 143

Operational aspects — *continued*
  simulation, 142, 143
  solvent interactions, 142, 143
  solvent losses, 142, 143
  solvent quality, 142, 143

Process
  definition, 2
  design, 2
Process control
  control points, 169, 170, 171, 172
  process control corrections, 171
  standard operating procedures, 171
Process development
  aims, constraints, 153
  bench unit, 154, 155
  chemical feasibility, 154, 155
  costing, 154, 155
  liquid–liquid contacting, 156
  liquid–liquid separation, 156
  means of transfer, 152
  mode of transfer, 152
  phase diagrams, 154, 155
  pilot unit, 154, 155
  separation factors, 154, 155
  technological design, 156
Process operation and requirements
  capacity, 82
  costs, 82
  mass transfer, 82
  phase separation, 82
  solubility, 82
  specificity, 82
Product
  distribution, 52
  quality, 4
  recovery, 76
  separation, 52

Reactions in LLX systems
  hydrolytic reactions, 130
  ion exchange, 137, 138, 139
  metathetic reactions, 34, 137

Second liquid phase
  aid to separation, 41
  biphase aqueous systems, 49, 166

bulk separation, 42
choice, 29, 42
fine separation, 42
in situ generation, 43
mechanism of transfer, 41
practical distribution coefficient, 43
recovery, 42
reversible transfer, 42
selection, 29, 42
separating/keeping separate, 32, 34
supplying reagent, 32, 34
transfer agent, 42
Separations
  analogy, 13
  definition, 1
  distribution coefficient, 96
  identify property for separation, 1
  main separation, 1
  multi operational, 99
  parameter sensitivity, 84
  practical stages, 102
  quality, 4, 43
  routes, 13
  secondary separation, 1
  separation factor, 96
  theoretical stages, 102
  transfer stages, 96
Solute
  characteristics, 1
Solvating solvents
  aliphatic primary alcohols, 101
  equilibrium curve, 101
  operating line, 102
  partial miscibility, 98
  ternary diagram, 101
Solvent
  acid extractant, 15, 16
  characteristics, 28
  choice, 2
  components, 15, 16, 24, 82
  composition, 82
  composition, testing, 96
  coupled transport, 74
  diluent, 15, 16, 24, 25, 82
  losses, 72, 84, 175, 176, 177
  modifier, 15, 16, 24, 25, 82
  polarity, 39
  reactant, 42
  selection, 18, 19, 20
  selectivity, 2, 4, 28

specificity, 2, 4, 28
stability, 84
transfer agent, 28
Solvent components
  active extractants, 1
  complexing agents, 1
  diluent, 1, 15
  modifier, 1, 15
  reagent, 1, 15
Solvent quality
  accumulations, 65, 145, 146, 167
  clean-up, 65, 167, 168
  deterioration, 65
  quality, 151, 174
  stability, 151
Solvent systems
  comparable conditions, 34, 35
  coupled transfer, 21
  interactions, 25, 105, 106
  modifications, 62, 63
  multiphase/multicomponent, 38
  systemization, 61
  transfer of material, 20
  transfer medium, 21
  types, 15, 16
Stripping
  pH control, 106
  salt formation, 106
  solvent addition, 106
System properties
  coalescence, 20
  dispersion, 20
  phase separation, 20
  significance, 20
System types, 104

Technological
  coalescence rates, 63, 163, 164, 165
  control
    locations, 65
    operational, 64, 105
    process, 64, 105
  engineering awareness, 105, 163, 164, 165
  entrainment, 163, 164, 165
  mass transfer rates, 63, 163, 164, 165
  nucleation, droplet growth, 166, 167
  phase contacting, 105, 163, 164, 165
  phase ratio, 105, 163, 164, 165

phase separation, 105, 163, 164, 165
steady state, 105, 163, 164, 165
transfer, 107, 108, 109
Technological feasibility
  coalescence, 151
  equipment selection, 151
  laboratory simulation, 7, 8, 9, 150
  mass transfer, 151
  scale-up, 149
  verification, 145, 146, 149
Test procedures
  counter-current, 32, 93, 115, 150
  cross-current, 32, 93, 115, 150
  modular multi-stage bench, 102, 115
  simulated counter-current-multiple batch, 102, 115
  single stage limiting conditions, 32
Test program
  minimum program, 118, 119
  interdisciplinary study, 77, 103
  process study, 93, 96, 98, 99, 100, 101, 102
  study project, 75
Thermodynamic properties
  activities, 84
  constants, 84
  scales, 84
Third-phase formation
  background material, 103
  reagent–solute-rich phase, 103
  representative case, 103, 104
  solid phase, 103
  technological approach, 104, 105, 166, 167
Transfer
  coupled transport, 74
  driving force, 91, 144
  from liquid to liquid, 107
  modes, 56, 57, 130, 131, 132, 133
  practical equilibrium constants, 107
  stages, 108
Two-liquid phase—*in situ* generation
  bipolymer/aqueous systems, 44
  in situ generation, 44, 51
  in situ self-generation, 43, 52
  polymer/salt/aqueous, 44
  water desalination, 44

Viscosity, 22

Water
  activity, 38
  balance (solvent-free basis), 72, 107
  distribution, 38
  LLX, 93
  mutual miscibility, 106
  partial vapor pressure, 107, 120